I Want To Be a Veterinarian...

I0661335

Table of Contents

I Want To Be a Veterinarian...

By
Brett Shayler

Written by

Brett Shayler

Book Cover by Karen Shayler

To refine the editing and book cover design, this book has used several artificial intelligence (AI) programs, such as ProWritingAid, Grammarly, QuillBot, Canva, and Adobe Photoshop.

Forward by Dr. Jennifer Bouche, DVM, CVA, CVC

When a young person tells me they want to be a veterinarian, it is clear that a deep compassion for animals has instigated their pursuit for a long, empathetic apprenticeship. This book captures all sides of this promise with a tenderness and honesty that will resonate deeply with older teen and young adult readers.

"I Want To Be A Veterinarian" follows a young woman whose dream of becoming a veterinarian is not a neat, romantic notion but a lived, sometimes messy pursuit. The author — whom I have had the pleasure of knowing first as a client and now friend— writes with an intimacy about the profession, the numerous avenues it can take, and the people and pets met along the way that help shape the journey. From the scenes in the exam room to the small triumphs and the quiet griefs ring true because they are informed not by cliché but by observation and heart.

What I value most in this story is its refusal to simplify what it means to be a veterinarian. The profession is equal parts science, patience, and stubborn empathy. It asks us to sit with uncertainty, to find courage in the routine, and to choose kindness even when choices are hard. The protagonist's struggles and growth can be familiar to those who have learned their craft at the side of another living being. I feel her sense of determination will be an encouragement to readers who are still finding their own path.

To young readers dreaming of following her footsteps: this novel shows you what the profession can ask of you, and what it can give back — deep connection, unexpected wisdom, and a sense of purpose. And to the new author: thank you for turning your observations and compassion into a story that will, I hope, inspire the next generation of animal lovers and caregivers.

As one of my mentors told me during my journey, "You can have everything, but not all at once."

Jennifer Boeche, DVM, CVA, CVC

Dedication

Idedicate this book to every young dreamer who dares to chase their passions, big or small. It is for the children who find solace and strength in the unwavering companionship of an animal friend, a bond that transcends words and fosters a love that lasts a lifetime. This story primarily celebrates those who show the quiet determination needed to overcome obstacles, persevere through setbacks, and believe in their potential, supported by a loyal companion.

It's for the children who devote countless hours to honing their skills, whether mastering a difficult equestrian move, perfecting a musical piece, or constructing an elaborate Lego castle. It's for those who recognized the value of commitment from the very start. Mornings, late nights, sweat, and tears contribute to achieving a goal for the children who know that the journey, with all its difficulties, is as important, if not more so, than the ultimate triumph. It's a journey made easier and more rewarding with the unwavering support of friends, family, and animal companions, a reminder that dedication and support are key to success.

This book also celebrates the incredible bond between humans and animals—a relationship built on trust, respect, understanding, and a shared love for the beauty and power of the natural world. It acknowledges the patience, care, and consistent attention that went into developing this significant connection. This story is for those who have experienced the magic of connecting with animals of all sizes, felt the shared moments of calm, and enjoyed the exhilaration of a decisive run. May the following pages inspire you to reach for your dreams, nurture your friendships, and cherish the bonds that enrich your life. And always remember that anything is possible with hard work, dedication, and belief in yourself.

The narrative within this book chronicles the journey of someone who becomes a veterinarian. The remaining books will explore in greater detail the distinct obstacles and victories encountered by young equestrians and their devoted companions. These stories will celebrate the unbreakable bond between people and animals, from the excitement of competing to the peaceful moments of connection, while also showcasing the lessons learned, and the friendships made during their time together.

Parents, guardians, and mentors are the intended audience of this book, which aims to assist and motivate young individuals. Your steadfast confidence in their capabilities, enduring patience, and devoted commitment to fostering their interests were the foundation of their achievements. We appreciate your being their source of strength, their guiding light, and their motivation. The love and support you give make a meaningful difference in their journey.

We also dedicate this book to the animals that constantly inspire us with their grace, strength, and steadfast loyalty, as a gesture of utmost appreciation. The profound impact you have on our lives is something that words cannot express. By teaching us the value of trust, the importance of patience, and the joy of shared adventures, you have made a profound impact on us. We always hope to hold in high regard and treasure the remarkable connection we have with you.

This book is for parents, guardians, and mentors who support and encourage young people passionate about helping animals. Your consistent faith in them, your capacity for patience, and your devoted commitment to their development influence the success. To discover additional details regarding the veterinary profession, please find out more information on how to become a veterinarian. We would like to encourage you to explore the resources available through the American Veterinary Medical Association and your state veterinary board office. To help you with future veterinary programs, we have provided a link to the association's website, which is www.AVMA.org. We extend our gratitude for your guidance, support, and the inspiration you have provided.

Chapter 1: Early Influences: The Family Farm

As I sit here, listening to the commencement speeches and hearing my classmates' names called as they prepare to receive their Doctor of Veterinary Medicine degrees, I can't help but reflect on how my journey began. My name is Samantha Green, and soon, I'll be Dr. Samantha Green, but more on that later. For now, I'd like to take you along on my path and share some of my earliest thoughts about it.

The scent of hay and the lowing of cows were the first things I remember. My earliest memories are a kaleidoscope of sights and sounds: the warm, earthy smell of the barn, the rhythmic clopping of hooves on the wooden floor, the soft bleating of lambs, and the comforting weight of a sleeping dog nestled at my feet. Our family farm, nestled in the heart of rolling hills, was more than just a place; it was a living entity — a vibrant ecosystem teeming with life. It was my sanctuary, my classroom, and the crucible where my passion for veterinary science was first forged.

My father, a man of quiet strength and unwavering dedication, instilled in me a deep respect for the land and its creatures. I learned the rhythms of the farm from him, the meticulous routines of milking, feeding, and tending to the animals. He was more than a farmer; he'd often spend hours with a sick or injured animal, his hands gentle, his touch reassuring, coaxing them back to health with a patience and understanding that I would later come to emulate in my veterinary practice.

My mother, a whirlwind of energy and boundless optimism, played a different, yet equally vital, role in shaping my early understanding of animal care. While my father focused on the practical aspects of farming, my mother infused our lives with an ache that felt strangely familiar and with compassion. She would tirelessly tend to our menagerie of pets–a motley crew that included dogs, cats, rabbits, chickens, and even a rescued goose with a broken wing–teaching me to approach each animal with sensitivity and respect. She nurtured our connection with the animals, emphasizing the importance of recognizing their personalities and needs.

One of my earliest memories involves Bess, our old, trusty mare. Bess had been with us for years; her coat was the color of sunbaked earth, and her eyes bore the wisdom of countless sunrises and sunsets. One evening, Bess went lame. A hesitant limp replaced her usual spirited gait, and pain clouded her usually bright eyes. My father, after a thorough examination, determined that she had a severe case of founder–a painful inflammation of the hoof.

I remember that night vividly. The air in the barn was thick with the scent of hay and Bess's anxious breaths. My father worked tirelessly, his brow furrowed in concentration as he carefully cleaned and treated Bess's injured hoof. My mother, ever the source of comfort and strength, kept a vigil by their side. I, though too young to offer any real help, sat quietly, absorbing every detail of their ministrations. It was in that dimly lit barn, surrounded by the familiar smells and sounds of the farm, that I first experienced a powerful urge to ease the suffering of another being.

We worked tirelessly through the night and the days that followed, implementing a regimen of poultices, rest, and carefully regulated exercise. My father's expertise, combined with my mother's unending care and Bess's resilience, led to a gradual recovery. Although a slow and painstaking process, Bess's ability to finally walk without a limp felt like a triumph.

The farm itself was a character in my childhood narrative. Its sprawling fields of golden wheat, punctuated by the occasional oak tree, provided an idyllic backdrop to our lives. The rushing creek, a silver ribbon weaving its way through the landscape, was where I would spend hours observing dragonflies flitting above the water's surface, frogs leaping from lily pads, and trout darting through the clear currents. Even the mundane tasks of the farm–collecting eggs from the henhouse, tending to the vegetable garden, or feeding the pigs–were opportunities to learn and connect with nature.

But the farm was not without its challenges. Winter storms brought with them the harsh realities of rural life–blizzards that buried our home in snowdrifts, icy winds that chilled us to the bone, and the ever-present struggle to ensure the well-being of our animals during the harsh winter months. These experiences, while often demanding and stressful, taught me the importance of resilience, resourcefulness, and the necessity of working collaboratively to overcome adversity.

The animals on the farm became my companions, my teachers, and my confidantes. There was old Tom, our grumpy but loyal Border Collie, whose unwavering loyalty and companionship taught me the power of unconditional love. Then there was Patches, our calico cat, whose feline grace and independence taught me the importance of observing and respecting the unique characteristics of each creature. There were the gentle cows, their eyes filled with quiet wisdom, and the playful lambs, their antics a constant source of amusement.

I remember vividly the day a lamb got separated from the flock. Panic set in as we scoured the fields, the icy wind whipping at our faces. I remember the adrenaline pumping through my veins as I searched for the lost creature, the feeling of relief that washed over me when we finally found him, cold, scared, and huddled under a bush. It was an experience that cemented in my heart the importance of compassion, responsibility, and the urgent need to protect vulnerable creatures.

Beyond the everyday routines of farm life, there were moments of wonder and awe. Witnessing a newborn calf take its first tentative steps, the sight of a mother hen protectively sheltering her chicks under her wings, and the majestic flight of a hawk circling overhead–these were experiences that nurtured a deep appreciation for the beauty and complexity of the natural world.

It wasn't just the animals that shaped my early life; it was the entire fabric of life on the farm–the changing seasons, the harsh realities of weather, the symbiotic relationships between the different organisms that thrived within its ecosystem. This immersion in the natural world instilled in me a deep-seated love for animals, a curiosity about their biology and behavior, and a desire to understand and ease their suffering.

The farm, with its beauty, challenges, and the inhabitants that thrived within its environment, provided the perfect foundation for my future aspirations. It was on that family farm, surrounded by the sounds of nature, the smell of hay and earth, and the company of animals both big and small, that the seeds of my dream were planted—a dream that would ultimately lead me to pursue a Doctorate in Veterinary Medicine, a profession dedicated to caring for the creatures I loved and respected. Throughout my life, the lessons, experiences, and relationships from that farm would shape my views and decisions.

The harsh realities of rural life often disrupted the idyllic setting, yet it was precisely these challenges that shaped my character and strengthened my resilience. The farm became my foundation, teaching me the value of hard work, dedication, and the profound rewards of caring for and healing animals. It was a life-changing experience that offered lessons I would carry with me forever.

Chapter 2: Discovering Veterinary Science: The First Competition

The county fair buzzed with activity, a kaleidoscope of sights and sounds that both thrilled and intimidated me. The air hummed with the energy of a thousand people, a blend of the sweet scent of cotton candy, the earthy aroma of livestock, and the underlying thrill of competition. This wasn't just any county fair; this was the stage for the regional Veterinary Science competition, and my heart hammered against my ribs like a trapped bird.

My team, the "Barn Buddies," comprised four other teenagers–Sarah, the pragmatic leader with an encyclopedic knowledge of bovine diseases; Mark, the quiet observer who possessed an uncanny ability to diagnose ailments with a single touch; Emily, a whirlwind of nervous energy and boundless enthusiasm; and David, the gentle giant with a calming presence that soothes even the most stressed-out animals. We were an unlikely group, bound by our shared passion for veterinary science and the daunting task before us.

As the youngest member, a knot of anxiety twisted in my stomach. Unlike my teammates who had competed before, this was my first rodeo. The weight of expectation from both myself and the team pressed down, threatening to smother the excitement I should have felt. My hands, usually steady, trembled slightly as I adjusted my well-worn overalls. I'd spent weeks preparing–poring over textbooks, practicing palpation techniques on our farm animals, and memorizing the intricate details of various animal diseases. But despite the hours of study, a nagging fear persisted that I wasn't good enough, that I would let my team down.

A practical exam focusing on diagnosing and treating common animal ailments was the first event. The instructors assigned a set of scenarios, each featuring a unique animal and symptom. Heavy pressure filled the air. Calm and efficient, Sarah, our fearless leader, took control and directed us. Mark's insightful quietness revealed subtle details others overlooked, proving invaluable. Emily's unwavering enthusiasm kept our spirits up; her nervous energy was oddly infectious. David, with his gentle hands and reassuring demeanor, proved remarkably adept at calming the distressed animals we examined.

My role involved assisting with physical examinations, taking vital signs, and documenting our findings. While I felt a little out of my depth, my teammates were supportive, offering guidance and encouragement. Their faith in me bolstered my confidence, replacing the gnawing self-doubt with a growing sense of purpose.

One particular scenario tested our mettle. We received a young lamb exhibiting symptoms of severe dehydration and respiratory distress. My heart pounded in my chest as I carefully assessed the lamb's condition, my hands surprisingly steady as I checked its temperature, heart rate, and respiration. Mark, with his insightful observation, noted a subtle tremor in the lamb's legs, suggesting possible neurological involvement, a detail that I had initially overlooked.

We worked together as a well-oiled machine. Sarah quickly diagnosed the dehydration and directed us to fluid therapy. Emily efficiently prepared the equipment. David's soothing touch calmed the distressed lamb, facilitating the administration of fluids. I meticulously documented our findings to ensure the accuracy of our diagnosis and treatment plan. The sense of teamwork, the shared responsibility, and the collective effort to save the lamb–it was both exhilarating and deeply satisfying.

Even more challenging was the written exam. The questions were comprehensive, delving into a vast array of topics from animal anatomy and physiology to advanced veterinary pharmacology. I pored over the questions, my mind working in overdrive. Some questions I knew with absolute certainty. Others required more deduction, more piecing together of knowledge gained from hours of study. I relied on my teammates, bouncing ideas off them, and engaging in collaborative problem-solving. Despite my initial apprehension, the shared struggle brought us closer, solidifying our bond.

As the day progressed, a sense of camaraderie grew among us. The initial nervousness gave way to a sense of focused determination. We had faced setbacks, stumbled, even felt the sting of doubt. But with each challenge overcome, our confidence grew, and our teamwork strengthened. The competition was no longer just about winning; it was about proving ourselves, about pushing our limits, about experiencing the thrill of collaborative effort.

The final judging felt surreal. We presented our findings to a panel of veterinary experts, our voices steady, our explanations clear. As we answered the judges' questions, I felt a surge of pride in my team, our collective knowledge, and the collaborative effort we had poured into the competition. We had worked together, learned from each other, and, most importantly, treated each animal with the utmost respect and care.

Later that evening, someone announced the results. You could feel the tension hanging like mist, a mixture of anticipation and anxious waiting. When they announced our team as the winner, a wave of elation washed over me. It was more than just a victory; it was a testament to our teamwork, our dedication, and our unwavering commitment to veterinary science. We had overcome our anxieties, embraced the challenge, and emerged victorious. The experience was a crucial step in my journey towards achieving my lifelong dream.

The seeds of my dream, planted on the family farm, had sprouted, grown, and blossomed under the pressure and excitement of that first competition. It was the first of many milestones, a prelude to the future successes to come, but more importantly, a confirmation of my chosen path, the unshakeable certainty that I was in the right place, doing the thing I did. Palpable excitement lingered, the taste of success both sweet and exhilarating, the reward deeply humbling and satisfying. My passion for veterinary medicine solidified through this experience, which also highlighted the collaborative efforts crucial for success and strengthened my bonds with my teammates, forging a lasting friendship beyond the awards. The journey was only beginning.

Chapter 3: The Thrill of Victory: Early Successes

The following year's regional competition felt different. My butterflies returned, but now a surge of confidence from our previous victory joined them. We, the Barn Buddies, were no longer the underdogs. We were the team to beat. The pressure was higher, the stakes seemingly greater, but this time, a determined focus overshadowed the fear.

Our preparation was meticulous. We spent countless hours dissecting cadavers, studying anatomical diagrams, and refining our practical skills. Sarah, ever the pragmatist, organized our study sessions with military precision, dividing responsibilities strategically. Mark's quiet observations were invaluable, often highlighting details others overlooked. Emily's energy remained contagious, lifting our spirits even during the most grueling sessions. And David, as always, served as our anchor, his calm demeanor radiating confidence and reassurance. This year, I felt more prepared. My role expanded, and my contributions felt more substantial. I took the lead in several diagnoses, drawing on the experience gained from the previous year's competition and from assisting Dr. Evans at the local animal clinic during my summer internship. I learned to trust my instincts, to rely on my knowledge, and to share my insights with my team confidently. My hands, once trembling with anxiety, now moved with practiced efficiency.

The competition itself was a whirlwind of activity. The practical exam was more challenging, the scenarios more complex, and the animals more difficult. We encountered a horse with a mysterious lameness, a cat suffering from acute respiratory distress, and a dog exhibiting neurological symptoms. Each case required a multifaceted approach, a coordinated effort that leveraged each member's unique skills and expertise. We debated, we questioned, and we collaborated; in each instance, we emerged victorious.

The written exam was equally demanding. The questions were intricate, requiring a deep understanding of animal physiology, pharmacology, and pathology. This year, however, I felt more prepared, more confident. The relentless hours of study, combined with the mentorship of Dr. Evans and the experience gained from the previous competition, had all contributed to my growing expertise. I approached each question methodically, carefully analyzing the facts, applying my knowledge, and planning a reasoned response. Again, teamwork was key. We checked each other's answers, reviewed our reasoning, and ensured that our responses were comprehensive and well-supported.

One particular moment stands out. We faced a tricky scenario involving a dog exhibiting unusual behavioral changes. Initial assessments pointed towards several diagnoses, each with vastly different treatment implications. We debated the merits of each possibility, each of us contributing our insights and observations. Mark, with his keen observational skills, noted a subtle change in the dog's gait that had previously gone unnoticed by us. This minor detail, coupled with my thorough review of the dog's medical history, led us to the correct diagnosis—a rare neurological condition. Solving this complex case felt like a profound personal accomplishment. The profound relief we felt for the dog matched the feeling of intellectual satisfaction.

Announcing the winners concluded the competition. You could feel the tension. The silence was intense. A second wave of euphoria, stronger than the first, hit me when they announced us. We didn't merely repeat a previous victory; it showed the power of collaboration, the strength of individual expertise combined with collective effort.

The celebrations were exuberant. We exchanged high fives, hugs, and joyous laughter. The bonds we shared were stronger than ever, forged in the crucible of competition and cemented by shared success. The joy was not solely about the trophies we had earned, but about the journey, the growth, and the enduring friendship we had cultivated. Each member of the Barn Buddies had contributed unique talents and perspectives, and the collective efforts of our collaboration had yielded remarkable results. The experience had reinforced the importance of teamwork, highlighting how mutual support and encouragement could transform individual strengths into a powerful force capable of achieving seemingly impossible goals.

The regional victory opened doors to new opportunities, including an invitation to the state-level competition, a much larger and more prestigious event. The scale was intimidating, but our previous successes had given us confidence. We approached the state contest with renewed purpose, driven not just to win but to test ourselves against the best in the state. The challenges were intense, and the level of competition far exceeded anything we had faced before. The practical exams grew more complex, and the written tests more demanding. We confronted unfamiliar challenges, including exotic animal diseases, advanced surgical techniques, and intricate diagnostic procedures. The pressure was intense, but we faced it together, backed by thorough preparation and seamless teamwork.

In the state-level competition, we faced challenges to our skills, refined our teamwork, and put our commitment to the test. The competition brought us closer, strengthening our bonds and reinforcing the importance of collaboration and mutual support. Each victory, each success, was a collective achievement, a testament to the power of teamwork, the strength of friendship, and the unwavering dedication that we all shared for veterinary science. It was also a personal triumph, a validation of years of hard work, commitment, and unyielding passion.

The last day of the state competition arrived, culminating in a presentation of our findings to a panel of distinguished judges. Our presentation was comprehensive, our responses articulate, and our teamwork clear. The atmosphere was electric; the tension was almost unbearable. When the judges announced the results and called our team the winners, relief and exhilaration washed over us. This time, the victory felt even sweeter, more profound.

Our collective efforts, dedication, and unwavering commitment to veterinary science resulted in a landmark achievement. Winning wasn't the only goal; it was about the shared journey, the laughter, the tears, the late-night study sessions, and the unwavering support we offered each other throughout our shared experience. This victory was more than just a trophy; it was a symbol of our uncompromising dedication, our shared passion, and the enduring bonds of friendship that we had forged on the road to achieving our collective dreams.

Our victory propelled us further along our paths, solidifying our commitment to veterinary science and serving as a stepping stone. Victory's thrill intoxicated us, fueling ambitions and solidifying our resolve to pursue dreams with unwavering determination and passionate enthusiasm. The road ahead remained long, but we faced it with renewed energy, reinforced by this significant accomplishment, and ready to take on whatever challenges lay ahead.

Chapter 4: The Sting of Defeat: Learning from Setbacks

Our years of toil and commitment culminated in winning the state competition, which felt like the ultimate achievement. That feeling of euphoria didn't last long, though. Compared to the state competition, the national competition was a monumental and daunting challenge. Though confident and proud of our achievements, we felt nervous as we reached Washington, D.C. This was more than just typical career advancement;

The national competition's immense size was overwhelming. Teams from across the country, each boasting exceptional talent and years of experience, filled the competition halls. The air buzzed with a palpable energy, a mix of anticipation, nervousness, and fierce determination. Suddenly, our previous victories felt less like achievements and more like stepping stones leading to this ultimate test.

The preliminary rounds were a blur of intense activity. The practical exams presented scenarios of unprecedented complexity, pushing our skills to their absolute limits. We encountered exotic animal diseases we'd only read about in textbooks, demanding immediate diagnosis and treatment. We faced advanced surgical procedures that required precision and swift decision-making under immense pressure. The written examinations were equally challenging, delving into intricate aspects of veterinary science that demanded a comprehensive understanding of the subject.

Our team faced its first significant challenge during the necropsy portion of the practical exam. A rare avian species exhibiting atypical symptoms challenged us with several dead ends in our initial diagnoses. We were under the clock; the pressure mounting with each passing minute. The initial confidence we had carried with us eroded, giving way to a growing sense of unease. Usually, our collective brainstorming sessions were a source of strength, a place where our strengths combined to overcome any obstacle. This time, however, the unusual nature seemed to stifle our usual dynamic.

Disagreements surfaced, something that had been almost unheard of before. Frustration and uncertainty disrupted the usual harmony. Arguments arose over diagnostic strategies, treatment plans, and even the interpretation of basic anatomical findings. Each disagreement chipped away at our team's cohesion. The usual seamless collaboration — our secret weapon — seemed to have fractured. The pressure of the competition was amplifying our weaknesses, making it challenging to leverage our collective strengths.

Heightened stress and palpable tension resulted from the lack of decisive progress among us. Each team member's doubts and anxieties became contagious, escalating the discord and hindering our collective efforts. We tried to maintain our calm, but time was slipping away. Finally, just minutes before the deadline, we reached a consensus — a decision that felt more like a compromise than a definitive outcome. We completed the exam, but the experience had left us shaken.

The sting of defeat was immediate and profound. Our team, once a seemingly unstoppable force, failed to place even among the top ten. Disappointment weighed heavily on us. The silence in our hotel room that night spoke volumes. The air was thick with disappointment, each member consumed by their thoughts, the team dynamic completely disrupted. We hadn't just lost a competition; we had lost our sense of invincibility, our shared confidence.

The following days were a period of intense introspection and self-reflection. The initial shock gave way to a wave of disappointment, a profound questioning of our abilities and the effectiveness of our teamwork. For the first time, we openly acknowledged our shortcomings.

Sarah admitted her leadership style may have become too rigid under pressure. Mark confessed that the unusual nature of the challenges presented had clouded his typically insightful observations. Emily's usual contagious energy waned, replaced by self-doubt. Even David, our ever-reliable anchor, acknowledged feeling overwhelmed.

Our post-mortem analysis was far more than a simple review of missed answers; it became a thorough analysis of our strengths and weaknesses, both individually and collectively. Our meticulous review encompassed diagnostic and treatment approaches, communication strategies, and decision-making processes. We identified crucial communication breakdowns, misinterpretations, and missed opportunities for collaboration. We recognized areas where we failed to use our strengths effectively and where competitive pressure highlighted our weaknesses.

Pressure can negatively affect even the most successful teams, transforming their collaborative energy into self-doubt and conflict. In retrospect, we did not fully grasp how crucial clear and consistent communication is, especially in times of high pressure and urgency. Upon reflection, we realized our team had not effectively leveraged its key strengths when facing the considerable pressures of the situation. To improve our teamwork, we had to create and implement strategies focused on enhancing communication channels, mitigating stress levels among team members, and maximizing the effectiveness of our collaborative efforts.

Seeking feedback was also crucial in our recovery process. We reached out to Dr. Evans, our mentor, for her insightful perspective. Her feedback was invaluable, not only identifying areas where we could improve our technical skills but also pinpointing crucial areas in teamwork that required strengthening. She highlighted the importance of proactively addressing disagreements and the need for a more adaptable approach to problem-solving. She stressed the importance of developing coping mechanisms for stress, ensuring that we could perform under pressure without compromising our collaborative spirit.

The road to recovery was long and arduous, but our shared disappointment proved to be a powerful catalyst for growth. Under simulated pressure, we practiced navigating difficult situations. Focusing on weak points, we sharpened our skills and enhanced our team's capabilities. Acknowledging mistakes as part of growth, we renewed our mutual trust. To maximize individual strengths, we restructured our roles. We included mindfulness techniques to improve concentration and calmness under duress.

The subsequent year saw us return to the national competition, not with the same carefree confidence as before, but with a tempered determination and a renewed appreciation for the challenges ahead. This time, our preparation was more meticulous, our communication more strategic, and our teamwork more resilient. We used the sting of defeat as fuel, transforming a setback into an opportunity for profound growth and development.

The national competition that year was not a breeze; we still encountered challenges, but our responses were more strategic and practical. Our communication was more transparent, our collaboration was smoother, and our problem-solving approaches were more effective. This time, the pressure didn't create discord; instead, it strengthened our resolve and sharpened our ability to perform. Our previous failure had forged a more resilient teamwork.

The outcome for this time was different. The trophy felt heavier, more significant, and more deeply earned. Our victory this time was not just a testament to our skills and knowledge, but also to our resilience, self-awareness, and the enduring strength of our team. We had transformed a painful setback into a catalyst for self-improvement, reinforcing the understanding that actual growth comes not only from our achievements, but also from our mistakes. The sting of defeat had ultimately strengthened us, wiser and more resilient. It had instilled in us a deeper appreciation for the value of teamwork, self-reflection, and the power of learning from setbacks. And that was perhaps a victory even greater than the competition itself.

Chapter 5: Mentorship and Guidance: Finding Support

The post-competition introspection, while painful, had laid bare the cracks in our foundation. We'd identified weaknesses in our communication and leadership, but fixing them required more than just self-reflection. We needed guidance, a seasoned perspective to help us navigate the complex terrain of teamwork and high-pressure performance. That's where Dr. Evans came in.

Dr. Evans wasn't just our veterinary science teacher; she was a mentor, a guiding light in the sometimes-turbulent waters of our ambitions. She possessed a rare blend of scientific expertise and empathetic understanding, a quality that made her an invaluable resource throughout our journey. Her office, usually bustling with students, became our sanctuary, a place where we could dissect our failures without judgment and rebuild our strategies with her expert guidance.

Our first meeting after the national competition was tense. We sat around her desk, a palpable silence hanging in the air, the weight of our defeat heavy on our shoulders. Dr. Evans, however, didn't offer platitudes or empty reassurances. Instead, she listened patiently, her keen eyes observing our every expression, absorbing the raw emotions we poured out. She allowed us to air our grievances, to vent our frustrations, and to acknowledge our shortcomings without interruption. Only after we'd exhausted ourselves did she speak.

Her words were not a condemnation, but a carefully constructed roadmap for improvement. While highlighting our mistakes with constructive, almost clinical precision, she didn't shy away. She pointed out the specific communication breakdowns, the missed opportunities for collaboration, and the points where individual anxieties had undermined our collective efforts. She meticulously analyzed our necropsy performance, revealing how differing interpretations caused delays and an incorrect diagnosis.

"You possessed all the knowledge," she stated, her voice calm and measured, "but under pressure, your communication faltered. Your strengths became weaknesses because you failed to leverage them collectively. You need to cultivate a more flexible approach to problem-solving, one that embraces diverse perspectives and uses each team member's unique skills."

Her guidance went beyond technical skills. Focusing on the psychology of teamwork, she highlighted the importance of emotional intelligence in high-pressure situations. Empathetic communication and simple explanations can mitigate stress's amplification of existing weaknesses, as she explained. She introduced us to stress-management techniques, suggesting mindfulness exercises and strategies for effective communication under pressure. She even recommended workshops on conflict resolution and team dynamics.

Dr. Evans didn't just offer advice; she actively participated in our recovery process. She organized practice sessions that simulated the high-pressure environment of the national competition. These sessions weren't just about technical skills; they tested our communication, decision-making, and conflict-resolution abilities. She created scenarios that forced us to confront our weaknesses and develop new strategies for collaboration.

One particularly challenging practice session involved a complex case of equine colic. The simulated scenario required a multi-faceted approach, involving diagnostic imaging, blood analysis, and a delicate surgical procedure. As we worked through the simulation, tensions flared once again. Old patterns emerged, and our usual collaborative spirit seemed to evaporate under pressure.

Dr. Evans, observing our struggles, stepped in, not to offer solutions, but to guide us towards them. Encouraging us to pinpoint the root causes of our conflict and brainstorm collaborative solutions, she facilitated a discussion. She pushed us to articulate our concerns and perspectives, helping us to understand each other's points of view. She showed the power of active listening, modeling effective communication strategies in real-time. Through her guidance, we learned to identify and address conflicts constructively, transforming potential disruptions into opportunities for growth.

Beyond Dr. Evans, other mentors emerged as crucial figures in my journey. My parents, while not directly involved in veterinary science, provided unwavering emotional support. Their faith in my abilities instilled in me the confidence to persevere through challenging times. They didn't just celebrate my victories; they also offered solace during setbacks, reminding me that setbacks were inevitable steps on the path to success. Their unwavering belief in me was a constant source of strength, a bedrock on which I could build my resilience.

Similarly, my college professors played pivotal roles in my development. Professor Ramirez, with his encyclopedic knowledge of avian anatomy and physiology, was an invaluable resource, patiently guiding me through complex research projects. His meticulous attention to detail and unwavering commitment to accuracy set a high standard that I strived to emulate. Professor Chen emphasized the importance of critical thinking and independent research, encouraging me to question established norms and challenge conventional wisdom. He fostered my intellectual curiosity, pushing me to explore new frontiers in veterinary science.

These relationships were not merely transactional; they were deeply personal and profoundly impactful. Each mentor offered unique perspectives, shaped my development in distinct ways, and nurtured my passion for veterinary science. They challenged me, encouraged me, and guided me through the complexities of academia and the demanding world of veterinary medicine. Their influence extended far beyond technical expertise; they helped me to cultivate resilience, confidence, and a deep appreciation for collaborative learning.

Academic guidance was only one aspect of the mentorship I received; it was about learning to navigate challenges, both personal and professional, with grace and determination. It was about finding strength in vulnerability, seeking help when needed, and acknowledging that even the most successful individuals need support and guidance along their journey. The support I received wasn't simply a helping hand; it was the cornerstone of my success. It was the fertile ground in which my dreams took root and flourished. These mentorships taught lessons that extended beyond the laboratory and the classroom. The collaborative spirit instilled by these experiences, the ability to learn from setbacks, and the capacity for self-reflection were the genuine gifts that prepared me for the

Challenges that lie ahead. They were the hidden curriculum, the invaluable lessons that proved far more significant than any textbook or lecture. And they were the seeds from which my future success would blossom.

Chapter 6: Balancing Academics and Extracurriculars

The national veterinary science competition served as a stark wake-up call, exposing the precarious balance I'd been attempting to maintain between my academic pursuits and extracurricular involvement. The intensity of the competition, with its grueling hours of preparation and immense pressure to perform, had thrown my carefully constructed schedule into utter disarray. Late-night practices and early-morning team meetings constantly disrupted my meticulously planned study sessions.

Sleep became a luxury, a fleeting respite snatched between demanding lectures and equally demanding competitions. My grades, once a source of pride, slipped, a subtle yet ominous sign of the unsustainable pressure I was under. The realization hit me like a physical blow: I couldn't continue down this path. Something had to change. The pressure wasn't just affecting my grades; it was seeping into every aspect of my life.

My relationships with friends and family suffered, replaced by a pervasive sense of isolation and exhaustion. My once vibrant social life dwindled to a series of hurried interactions, punctuated by apologies for missed events and commitments. Even my passion for veterinary science, once a boundless source of energy and motivation, felt like a burden, a relentless cycle of stress and anxiety.

The first step towards rectifying this imbalance involved a brutally honest self-assessment. I meticulously analyzed my schedule, identifying areas of inefficiency and unnecessary time wastage. I learned to prioritize ruthlessly, focusing on the tasks that directly contributed to my academic goals and competition performance. This meant saying "no" to activities that, while enjoyable, were ultimately non-essential. It was a painful process, requiring self-discipline I hadn't yet mastered. But it was also liberating, freeing me from the overwhelming feeling of being constantly pulled in multiple directions.

Time management became an obsession, a critical skill I had to hone to survive. I experimented with various techniques, from the Pomodoro method to the Eisenhower Matrix, seeking the strategies that best suited my learning style and personality. I discovered the power of scheduling, blocking out specific time slots for studying, practicing, and even relaxing. This structured approach provided a much-needed sense of control, mitigating the chaos that had previously engulfed my life.

Beyond merely adjusting my schedule, I recognized the crucial importance of effective study habits. The cramming I had relied on before was utterly ineffective under the pressure of my demanding routine. Instead, I committed to studying consistently and intensively throughout the week. Regular self-testing solidified my understanding of complex concepts through the use of active recall. I also discovered the value of spaced repetition, reviewing material at increasing intervals to enhance long-term retention.

The support of my family and friends proved invaluable during this challenging period. My parents, witnessing the strain I was under, offered unwavering encouragement and practical support. They helped me manage my time, taking on some of the household responsibilities that had previously fallen on my shoulders. Their understanding and empathy provided a much-needed buffer, helping me navigate the stressful period without feeling entirely overwhelmed.

My friends, too, played a critical role in my well-being. They understood the pressures I faced, offering words of encouragement and understanding. They helped me maintain a sense of normalcy, reminding me of the importance of balance and self-care. Their friendship provided a much-needed respite, a welcome escape from the demanding world of academics and competitions.

Beyond the emotional support of my loved ones, I actively sought strategies for stress management. I discovered the power of mindfulness, practicing meditation and deep breathing exercises to calm my anxious mind. Regular physical exercise became an essential part of my routine, providing a much-needed release of tension and promoting both physical and mental well-being. I learned to prioritize sleep, recognizing its critical role in cognitive function and overall health. These self-care practices, initially viewed as luxuries, became indispensable elements of my success.

However, the most significant shift came from reevaluating my approach to the veterinary science competitions themselves. I realized that my relentless pursuit of perfection, while admirable, had become self-defeating. The intense pressure to succeed had overshadowed the joy and passion that had initially drawn me to veterinary science. I focused less on winning and more on the learning process itself, appreciating the valuable experiences and skills I gained from each competition, regardless of the outcome. This shift in perspective, while initially challenging, proved profoundly liberating, reducing the stress and anxiety that had previously consumed me. It allowed me to approach competitions with a more balanced perspective, appreciating both the challenge and the opportunity for growth.

The journey of balancing academics and extracurricular activities was far from easy. It required relentless self-discipline, strategic planning, and the unwavering support of my family and friends. It involved setbacks and moments of self-doubt, but ultimately, it was a transformative experience. I learned invaluable lessons about time management, prioritization, stress management, and the importance of self-care. These lessons, far from being confined to the realm of academics and competitions, proved invaluable in all aspects of my life. They laid the foundation for a more balanced and fulfilling existence, one that celebrates both personal growth and academic achievement.

The experience taught me that success wasn't just about reaching ambitious goals; it required a delicate balance, a constant dance between striving for excellence and caring for my well-being. It was a lesson that would stay with me, shaping my approach to challenges throughout my life and providing me with the tools I needed to succeed not only in veterinary science but in all endeavors. It was in this struggle for equilibrium that I discovered the true meaning of perseverance, the understanding that the journey itself, with all its challenges and triumphs, was as valuable as the final destination.

The process of balancing competing demands honed my abilities, strengthened my character, and ultimately prepared me for the even greater challenges that lay ahead on my path to becoming a veterinarian. The lessons were hard-won, but they would serve as invaluable assets, guiding me through every step of my journey. That demanding year left me scarred, but those scars became badges of honor, testaments to my resilience and dedication.

Chapter 7: Navigating Friendships and Relationships

The single-minded chase for something just out of reach in veterinary science had, undeniably, taken its toll. My once vibrant social life had shrunk to a shadow of its former self. Weekends, previously filled with laughter, spontaneous adventures, and the comforting presence of friends, were now consumed by study sessions, practice runs, and the gnawing anxiety of upcoming competitions. The guilt gnawed at me, a constant companion whispering accusations of neglect and selfishness. Sarah, my closest friend since elementary school, withstood my absence. We'd shared countless secrets, dreams, and late-night pizza binges. Now, hurried phone calls punctuated by apologies and promises of "catching up later"—promises I stubbornly failed to keep—replaced our shared moments.

The strain on our friendship was palpable. Her gentle reminders of our dwindling time together subtly hinted at her disappointment, cracking the foundation of our friendship. I saw the flicker of hurt in her eyes during our rare encounters, and it pierced me with a sharp, self-inflicted pain. I knew I was pushing her away, unintentionally sacrificing our years of friendship on the altar of my ambition. It was a heavy price to pay, a stark realization that brought me face-to-face with the collateral damage of my relentless pursuit. I realized the irony: my unwavering dedication to helping animals was inadvertently causing me to neglect my friends.

My family, too, felt the impact of my single-minded focus. Dinner conversations, once lively exchanges of shared experiences and laughter, became stilted affairs, interrupted by my frantic attempts to cram in last-minute studying. My younger brother, Mark, once my confidant and playmate, grew increasingly distant; quiet resentment replaced his playful banter. I felt the weight of their unspoken disappointment, the silent accusation in their eyes that I was prioritizing a dream over them. The guilt intensified, a heavy cloak that suffocated my joy and fueled a sense of isolation.

This period forced me to confront the painful truth: success, no matter how rewarding, could never justify sacrificing my relationships. I couldn't afford to lose the support system that had buoyed me through countless challenges. It was time to reevaluate my priorities and find a more sustainable balance between ambition and connection. The change had to be more than just a tactical change of my schedule; it needed to be a fundamental shift in my perspective.

The first step involved open communication. I sat down with Sarah, confessing my failings and expressing my deep regret about my negligence. The conversation proved difficult; But it was also immensely healing. Her forgiveness, offered with the unconditional love that only genuine friendship can provide, lifted a weight from my shoulders. It wasn't a magical fix, but it was a crucial first step toward rebuilding our bond. We established new boundaries, committing to regular, dedicated time together, even if it meant sacrificing a few hours of studying each week. This wasn't about equal time; it was about intent and presence.

Similar conversations with my family followed. I acknowledged their feelings, expressing my remorse for neglecting them and promising to reconnect with them. Offering practical solutions, my parents took on some of my household responsibilities, freeing up time for me to connect with them. My brother, initially hesitant, responded positively to my efforts; his quiet resentment gradually gave way to renewed warmth.

Focusing on my relationships again involved more than just recovery; I attempted to be present, to focus, and to engage wholeheartedly in our shared activities. We started having family game nights, something that had become a distant memory, filled with laughter and the warmth of togetherness. I started having regular coffee dates with Sarah, simply enjoying each other's company, free from the usual anxiety and pressure to catch up on lost time. These slight gestures, these moments of intentional connection, proved to be far more effective than any elaborate gestures of atonement. This experience taught me the true meaning of balance. It wasn't about dividing my time equally between academics and personal life; it was about integrating them, recognizing their interconnectedness. My academic success wasn't an isolated achievement; it was a testament to the unwavering support of my friends and family, the strength that stemmed from their love and belief in me.

The lessons of compromise and understanding extended beyond my immediate circle. My intense focus also affected my relationships with professors and teachers. Several late submissions and missed office hours highlighted my struggle to maintain balance. I found that honest communication, outlining my commitments and challenges, was crucial. My professors, understanding the demands of my extracurricular activities, provided extensions and extra help, showing an empathy that deepened my respect for them.

The road to equilibrium was a winding one, marked by occasional setbacks and moments of self-doubt. Sometimes the pressure threatened to overwhelm me again, tempting me to retreat into the familiar comfort of solitary study. But I had learned the hard way that genuine success wasn't solely about academic achievements; it was about nurturing the relationships that sustained me, about fostering a life rich in love, laughter, and connection. The support network I cultivated became my anchor, providing the stability and strength I needed to navigate the stormy seas of high school and beyond.

As I reflected on my life, it occurred to me that a strong correlation between my overall well-being and the health of my personal relationships — a realization that significantly affected my perspective. The root of my burnout was far more complex than simply excessive work hours; it involved a combination of factors. Recognizing the interconnectedness of academic success and healthy relationships, I prioritized tending to my emotional, mental, and physical health, understanding that this holistic approach was crucial for achieving my goals in both spheres of life.

As I navigated my final years of high school, my schedule remained demanding, but I was no longer consumed by it. I had learned to set boundaries, to prioritize my well-being, and to cherish the invaluable support system that sustained me. Though challenges persisted, I could now meet them with a stronger self. The lessons I learned weren't merely about effective time management or achieving academic success; they were about the critical importance of nurturing the human connections that give life its meaning, and the realization that actual achievement extends far beyond personal ambition. The harmony I found between my academic pursuits and personal life wasn't a perfect equilibrium, but a dynamic balance—a constant change, a lifelong journey of learning and growth. And it was a journey I wouldn't have survived without the unwavering support and understanding of those around me. This lesson, learned through sacrifice and reconciliation, would prove invaluable as I navigated the intense pressures of veterinary school.

Chapter 8: The Pressure Cooker National Competitions

T he regional competitions had been a grueling test of knowledge and skill, but the nationals were a whole different beast. The air crackled with an undercurrent of nervous energy, a silent battle of wills waged between hundreds of bright-eyed, fiercely competitive students. Each competitor carried the weight of months, even years, of preparation, their dreams distilled into a series of meticulously planned presentations, practical exams, and rapid-fire question-and-answer sessions. The sheer scale of the event was overwhelming; the vast convention center buzzed with a frenetic energy, a vibrant hive of activity punctuated by the nervous chatter of students and the hushed whispers of anxious parents.

My carefully constructed routine, my sanctuary of organized chaos, felt utterly inadequate in the face of this colossal challenge. The familiar comfort of my study habits disintegrated under the weight of the immense pressure. Sleep became a luxury I could barely afford, snatched in short, restless bursts between grueling practice sessions and frantic last-minute reviews. The vibrant colors of my carefully organized flashcards seemed to blur, their once-clear information melting into a hazy, overwhelming jumble. The confidence I'd painstakingly cultivated crumbled, replaced by a gnawing self-doubt that threatened to consume me.

Adrenaline and intense focus characterized: The competition's first day, which was a blur of adrenaline and intense focus. The written exam — a marathon of intricate questions — tested the limits of my endurance and knowledge. My carefully memorized facts and principles swam in a dizzying whirlpool of information, the pressure threatening to overwhelm my carefully honed techniques. I pushed myself, forcing my mind to grapple with complex concepts, fighting against the growing fatigue that threatened to paralyze me.

My hands trembled as I scribbled my answers, my heart pounding a frantic rhythm against my ribs.

The theoretical and practical exams proved equally challenging, significantly hindering many students. Because the procedures were so precise and delicate, the competition's suffocating pressure undermined the composure and precision. With the rhythmic beeping of the heart monitor, the judges' anxious gazes, and the nervous tension that radiated from my fellow competitors all swirling around me, I found myself in a pressure cooker of an environment that truly pushed me to my absolute limits. At certain points during the performance, my hands trembled, my movements became awkward and unsure — a significant difference from the fluid, self-assured skill I had painstakingly honed through countless hours of dedicated practice. I could feel the sweat beading on my forehead, my breath catching in my throat. I fought back the rising panic, drawing strength from years of training, hours of preparation, and an unwavering belief in my abilities.

The oral examination was perhaps the most daunting of all. Facing a panel of distinguished veterinarians, I had to show not only my knowledge but also my ability to think critically, to respond swiftly and decisively under pressure. Each question was a challenge, a test of my intellectual agility and composure. I could hear the subtle undertones in their questions, the subtle probing that sought to expose any gaps in my knowledge. I struggled to maintain eye contact, my tongue feeling thick and clumsy in my mouth. There were moments of agonizing silence, where I felt the weight of their scrutiny, the unspoken expectation that I would answer flawlessly.

Yet, amidst the intensity and pressure, there were moments of unexpected grace. The camaraderie among the competitors, initially unseen beneath the competitive veneer, unexpectedly emerged. We shared nervous glances, small smiles of mutual understanding, a silent acknowledgment of the shared experience, and the overwhelming pressure we were all enduring. There was a quiet strength in that shared vulnerability, a silent pact of mutual support in the face of an immense challenge. These brief moments of human connection provided unexpected respite, a momentary release from the relentless tension.

We spent evenings huddled in our hotel rooms, poring over notes and sharing snippets of the day's challenges, finding comfort in shared anxieties and nervous laughter. The unspoken support became a powerful antidote to the isolating pressure, transforming a potentially solitary struggle into a shared experience. These quiet moments of connection fostered a sense of camaraderie that transcended the competitive aspect of the event, reminding me we were all united in our shared passion for veterinary science.

The last day of the competition felt like a sprint to the finish line. The accumulated pressure of the previous days seemed to coalesce into a formidable force, threatening to overwhelm me with its intensity. Yet, a strange sense of calm settled over me as I approached each challenge, a quiet confidence born from the realization that I had given my all, that I had prepared as thoroughly as possible, and that the outcome was beyond my control.

The awards ceremony was a surreal blend of anticipation, nervous excitement, and quiet pride. Hearing my name announced among the finalists was a moment of euphoric disbelief. The recognition was a testament to the countless hours of practice, the unwavering support of my mentors and family, and the belief in my abilities.

Competing at the national level forged my resilience, sharpened my skills, and expanded my understanding of my capabilities. The pressure was immense; the stakes were high, but it was also a profoundly rewarding experience, an affirmation of my passion and a testament to the power of perseverance. It taught me the importance of meticulous preparation, the critical role of mental fortitude, and the unexpected value of finding connection and support within the competitive arena. This experience, while incredibly intense, would prove invaluable as I embarked on the even greater challenge of veterinary school. The lessons learned in the pressure cooker of national competitions–resilience, adaptability, and the power of human connection–would become my guiding stars, illuminating my path through the years of intense study and demanding work that lay ahead. The national competition was not just a test of knowledge and skill; it was a formative experience, a crucial step in my journey toward achieving my dream. And the memories, both exhilarating and terrifying, remain etched indelibly in my mind, serving as a constant reminder of the sacrifices, struggles, and ultimate triumphs along the path to becoming a veterinarian.

The competition also illuminated the importance of self-compassion. I had pushed myself relentlessly, often to the point of exhaustion and burnout. Learning to recognize my limits and prioritize self-care and rest became as crucial as mastering complex veterinary techniques. It wasn't a sign of weakness to acknowledge my need for rest and recuperation; it was a sign of wisdom and self-awareness, essential for sustained success. This lesson, deeply ingrained during those demanding weeks, would prove invaluable as I navigated the intense pressures of veterinary school and beyond.

In retrospect, the national competitions were about more than just winning or losing. They forced me to confront my vulnerabilities, to push beyond my perceived limitations, and to discover the surprising strength that lived within me. It was a journey of self-discovery, an awakening to my inner resilience, and an appreciation for the supportive network that fueled my journey. The echoes of those intense days still reverberate, a powerful reminder of the tenacity I discovered within myself—a tenacity that would carry me through the many challenges that lay ahead on the path to becoming a veterinarian.

Chapter 9: Unexpected Opportunities Expanding Horizons

The national veterinary science competition wasn't just about ribbons and accolades; it was a launchpad. While the pressure was immense, the experience opened doors I hadn't even known existed. One such opportunity materialized unexpectedly during the post-competition networking reception. Dr. Anya Sharma, a renowned veterinary oncologist and one of the competition judges, approached me. Her piercing gaze, initially intimidating, softened as she complimented my presentation on feline leukemia. It wasn't just polite flattery; she saw something in my work — a spark of genuine passion and meticulous research that went beyond textbook answers.

She invited me to spend a week shadowing her at the renowned City Veterinary Oncology Center. This wasn't merely a shadowing experience; it was a masterclass. Dr. Sharma didn't treat me like a student; she treated me as a colleague, albeit a junior one. I scrubbed in on complex surgeries, witnessed the delicate art of chemotherapy administration, and learned about the emotional toll of working with animals facing life-threatening illnesses. The experience was both exhilarating and humbling. I saw the best and worst of veterinary cancer treatment—the commitment, the sadness, and the deep satisfaction of caring for animals, even when things looked hopeless.

Beyond the technical skills, I gained a perspective that reshaped the way I viewed things, focusing on the human side of veterinary medicine. Dr. Sharma's compassion for her patients, both animal and human, was palpable. She had a unique ability to connect with clients, offering comfort and support during difficult times. She taught me that veterinary medicine wasn't just about technical expertise; it was about empathy, communication, and a deep understanding of the human-animal bond. The week at the oncology center solidified my commitment to veterinary science and significantly broadened my horizons. I discovered a newfound passion for oncology, a specialization I hadn't even considered before.

Another unexpected opportunity emerged during a presentation at a regional science fair. My research project on the efficacy of a novel antibiotic in treating bacterial infections in reptiles caught the attention of Professor David Miller, a renowned herpetologist and professor at the State University. He invited me to join his research lab during the summer. This was more than just a summer job; it was an immersion into the world of scientific research. I gained expertise in experimental design, data analysis, and scientific writing. I worked alongside graduate students, contributing to ongoing research projects and developing my laboratory skills. The experience wasn't always easy; there were long hours, frustrating setbacks, and the occasional feeling of being overwhelmed. But the rewards surpassed the challenges.

I learned to work independently, to think critically, and to persevere in the face of obstacles. I developed a deep appreciation for the scientific method, for the painstaking process of testing hypotheses, analyzing data, and drawing conclusions. This was a pivotal experience that shaped my future academic path, cultivating a passion for research that would ultimately lead me to pursue a doctoral degree. The meticulous nature of the study instilled a discipline that was invaluable throughout my subsequent academic pursuits. The failures and setbacks were equally valuable, teaching me resilience, the importance of rigorous data analysis, and the need to approach research with humility and a critical eye.

Professor Miller became a mentor, guiding me through the complexities of the research process and offering encouragement and support. His guidance extended beyond the lab; he helped me navigate the intricacies of college applications, offering valuable advice on selecting a university and preparing for the intense academic challenges that lay ahead. His mentorship was a pivotal aspect of my academic success, illustrating the profound impact of a supportive and experienced mentor on a student's journey. The confidence he fostered in my abilities provided the bedrock upon which I built my subsequent academic achievements.

Yet another unexpected opportunity emerged through my participation in a national student veterinary association conference. A representative from a renowned animal shelter in a neighboring state approached me after a presentation on animal welfare. Impressed by my knowledge and passion, she offered me a summer internship at their facility. This internship was a profound immersion into the practical aspects of veterinary care. I worked alongside experienced veterinary technicians, assisting in a wide range of procedures, from routine vaccinations to emergency surgeries. This was real-world veterinary medicine at its most challenging and rewarding.

The sheer variety of cases I encountered broadened my understanding of veterinary medicine beyond the confines of textbooks and classrooms. I learned to handle difficult situations, to work efficiently under pressure, and to make quick, informed decisions. The experience taught me the importance of teamwork, of collaboration, and of clear communication. It also exposed me to the emotional complexities of working with animals in need, strengthening my empathy and compassion for animals and their human companions. I witnessed firsthand the challenges of working in a high-volume shelter setting, where resources are often limited and the workload is immense. This experience was more valuable than any textbook could ever provide. It provided me with a deeper understanding of the practical challenges and rewards of veterinary work.

These unexpected opportunities–the shadowing experience with Dr. Sharma, the summer research project with Professor Miller, and the internship at the animal shelter–were pivotal in shaping my trajectory. Acquiring new skills or adding to my resume wasn't their sole purpose; networking, proactively seeking experiences, and openness to unexpected paths leading to incredible opportunities: that's what they emphasized to me.

Because these experiences connected, their cumulative impact exceeded the sum of their individual parts. My time spent in Dr. Sharma's oncology lab honing my skills proved invaluable during my animal shelter internship, where I could confidently perform complex procedures because of this training. Professor Miller's guidance honed my research skills, enabling me to assess research critically and keep up with advancements in veterinary medicine. These opportunities built a sound foundation for my future success.

My relationships with mentors and colleagues was equally beneficial. The support system I developed from these opportunities surpassed basic professional networking; Dr. Sharma, Professor Miller, and the staff at the animal shelter weren't just mentors; they were role models, demonstrating the dedication, compassion, and resilience necessary to succeed in the demanding field of veterinary medicine. Their constant support and faith in my capabilities inspired me and guided me through tough academic times.

These unexpected opportunities also highlighted the importance of perseverance and adaptability. Not every experience was smooth sailing; there were moments of frustration, setbacks, and self-doubt. But navigating those challenges, learning from my mistakes, and emerging stronger helped to build the resilience that proved crucial throughout my academic career. This wasn't just about achieving academic excellence; it was about developing as a person, becoming more confident, resourceful, and resilient.

The journey toward becoming a veterinarian wasn't a straight line, but a winding path, rich with unforeseen opportunities that profoundly shaped who I was to become. It was a testament to the power of embracing the unexpected, of being open to new experiences, and of recognizing the transformative power of seizing opportunities.

Chapter 10: Graduation and the Next Step: College Applications

The final bell of my senior year rang, not with a simple chime but with a resounding clang that echoed the culmination of years of relentless effort. High school, once a daunting prospect, had become a blur of late-night study sessions, exhilarating competitions, and unexpected opportunities that had shaped me in ways I couldn't have imagined. Now, the next chapter loomed—college applications, a daunting mountain range of essays, transcripts, and recommendations that threatened to overwhelm even the most organized student.

The sheer volume of applications felt paralyzing. I had compiled a list of schools, each with its unique allure: the prestigious Cornell University with its renowned veterinary program; the charming University of California, Davis, known for its pioneering research; and the innovative University of Pennsylvania, with its emphasis on hands-on learning. Each application felt like a miniature autobiography, demanding a distillation of my academic achievements, extracurricular activities, and personal aspirations into a concise and interesting narrative. The pressure was immense, a stark contrast to the relative simplicity of high school exams. This wasn't simply a test of knowledge; it was a test of my ability to articulate my passions and aspirations, to convince admissions committees that I was worthy of a place in their esteemed programs.

The application essays were challenging. Each prompt required introspection, forcing me to confront my strengths and weaknesses, as well as my successes and failures. One essay asked me to describe a pivotal moment that shaped my career aspirations. I naturally gravitated towards my experience at the City Veterinary Oncology Center, describing the emotional toll of working with animals facing terminal illnesses, but also the profound satisfaction of providing comfort and care even in the face of death. Another essay focused on my research project on reptile antibiotics, highlighting the challenges of experimental design, the setbacks I faced, and the perseverance that finally yielded meaningful results. Each word felt weighted with the significance of my future, every sentence a testament to my commitment to the veterinary profession.

The letters of recommendation presented their own set of challenges. I had to select individuals carefully who could genuinely speak about my abilities and character. Dr. Sharma, Professor Miller, and Ms. Evans, the director of the animal shelter, readily agreed, each offering their invaluable perspectives. I felt a surge of gratitude for their willingness to vouch for me, to put their reputations on the line to support my aspirations. Their letters, I hoped, would convey volumes about my character and capabilities, surpassing the dry statistics and grades that comprised the rest of my application. The preparation wasn't simply academic; it was personal, a process of self-reflection that allowed me to assess my strengths and acknowledge areas for growth critically and honestly.

Moments of both intense excitement and crippling self-doubt punctuated the application process. There were the sleepless nights spent perfecting essays, the anxious hours spent waiting for application portals to update, and the near-constant fear I hadn't done enough, that I wasn't good enough. There were countless hours spent poring over the details of each application, meticulously reviewing every word and sentence, to ensure that every aspect of my application reflected the entirety of my academic rigor, personal strengths, and unwavering commitment to the field of veterinary medicine.

Beyond the academic aspects of the applications, I grappled with the personal implications of this pivotal decision. The choice of university was more than just an academic one; it was a decision about my future, my life, my identity. Each university held a unique promise, a different path towards my goal of becoming a veterinarian. I imagined myself immersed in the bustling academic environment of Cornell, the laid-back atmosphere of Davis, or the dynamic energy of Penn. Each possibility carried its own weight of expectation and anticipation.

The waiting period between submitting the application and acceptance was agonizing. Each day felt like an eternity, the anticipation building to a near-unbearable level. The constant checking of email, the refreshing of online portals, the nervous pacing–these became my rituals, my attempts to control the uncontrollable. The uncertainty was an unwelcome guest that haunted my every waking moment. I found solace in the support of my friends and family, each conversation offering a distraction from the agonizing wait. Their encouragement and unwavering belief in my abilities provided a much-needed anchor during this turbulent period.

Then, on one crisp fall evening, amidst the changing leaves and the crisp fall air, the acceptance emails arrived. Each one felt like a minor victory, a testament to the years of hard work, the countless hours of study, and the unwavering dedication I had poured into my academic pursuits. Cornell, Davis, and Penn—all three prestigious universities — extended offers of admission. The initial euphoria quickly gave way to a new wave of challenges—the daunting task of choosing between three exceptional programs.

The decision-making process was complex, requiring careful consideration of various factors: the reputation of each program, the research opportunities available, the faculty involved, and the overall academic environment. Ultimately, I chose Cornell, drawn to its rigorous program, its commitment to research, and the inspiring faculty. It was a decision based not solely on reputation or prestige, but on a deep sense of alignment with the program's values and its potential to further shape my academic journey. My acceptance to Cornell marked not just the end of one chapter but the thrilling beginning of another, a new phase of learning, discovery, and growth.

The journey had been challenging, filled with its share of setbacks and self-doubt. But it had also been rewarding, teaching me resilience, perseverance, and the transformative power of pursuing one's dreams. The long, arduous process of applications was now behind me, replaced by the fresh, exhilarating energy of anticipation for the next step. The future stretched ahead, brimming with opportunities and challenges. Still, I approached it not with fear but with excitement, ready to embark on the next phase of my journey toward becoming a veterinarian. Now, the high school years, once a seemingly endless stretch of academic rigor, felt like a distant yet cherished memory.

Chapter 11: The Rigors of Veterinary School Anatomy and Physiology

The crisp fall air of Ithaca, New York, held a biting chill that mirrored the intensity of my first semester at Cornell University's College of Veterinary Medicine. The idyllic campus, a picturesque blend of ivy-covered buildings and sprawling green lawns, belied the brutal academic rigor that awaited within. High school, with its relative predictability and manageable workload, felt like a distant dream. Veterinary school was a different beast entirely — a relentless onslaught of complex information, demanding practical skills, and an unwavering commitment to excellence.

My first hurdle was anatomy. The sheer volume of material was overwhelming. Days blurred into nights as I navigated the intricate labyrinth of the canine skeletal system, meticulously tracing the articulation of each bone, memorizing the origins and insertions of muscles, and identifying the subtle nuances of ligamentous structures. Cadaver labs, though initially daunting, became strangely familiar spaces, transforming the cold, lifeless forms into intricate puzzles begging to be solved. The pungent formaldehyde scent, initially repulsive, gradually became the olfactory marker of intense focus and intellectual engagement. I spent hours dissecting, identifying, and labeling structures, each successful identification a minor victory against the overwhelming tide of information. The intricate detail of the circulatory system, the delicate complexity of the nervous system, the functional elegance of the digestive tract—each system revealed a level of complexity that both challenged and captivated me. I spent late nights poring over Gray's Anatomy, tracing pathways and memorizing Latin terms, while our study groups consumed weekends; our collective groans and sighs testified to our shared struggle.

Physiology, the study of how the body functions, was equally demanding. The intricate dance of hormones, the complex interactions of organ systems, the precise mechanisms of cellular respiration—it all felt like an elaborate clockwork mechanism, requiring a deep understanding of each component to appreciate the functioning of the whole. We spent countless hours delving into the intricacies of cardiovascular dynamics, respiratory mechanics, and renal physiology, working through complex equations and interpreting physiological graphs to gain a deeper understanding. Each lecture felt like a race against time, a frantic scribbling of notes to capture the torrent of information pouring from the professor's lips. My whiteboard, covered in a chaotic tapestry of diagrams and equations, became a testament to the tireless effort required to master these concepts. I recall one particularly challenging assignment, which involved calculating the glomerular filtration rate and renal blood flow in various scenarios. The sheer complexity of the calculations and the intricate interplay of variables pushed me to the limits of my mathematical abilities. Success, however, came not from rote memorization but from cultivating a deeper understanding of the underlying principles.

The relentless pace of veterinary school extended beyond lectures and labs. Clinical rotations provided invaluable practical experience, but also brought their own set of challenges. The unpredictable nature of veterinary medicine demanded quick thinking, decisive action, and an unwavering commitment to animal welfare. The emotional toll was significant. Dealing with critically ill animals, witnessing the suffering of beloved pets, and communicating complex diagnoses to grieving owners demanded an emotional resilience I hadn't yet developed. I learned to balance empathy with professionalism, to find strength in moments of vulnerability, and to find solace in the simple act of providing comfort and care, even in the face of death. I learned to be an advocate, not only for the animals in my care but also for their distraught owners, guiding them through the emotional minefield of loss and grief.

Fortunately, I wasn't alone on this arduous journey. My classmates became my support system; our shared struggles to forge bonds of camaraderie and mutual respect. Study groups became havens of shared learning, late-night coffee sessions fueled by shared anxieties and mutual encouragement. We celebrated each other's successes and commiserated over setbacks, creating a culture of collaborative learning and unwavering support. The faculty, too, played a significant role in my success. Because the course was so demanding, many professors provided extra support and guidance for students. They were not just teachers, but mentors who guided me through the complexities of the curriculum and offered invaluable advice and encouragement. Their dedication to their students was truly inspiring, reflecting a passion for teaching that rivaled their passion for veterinary medicine. Their mentorship extended beyond the classroom, often into late-night emails, where they answered questions or provided a word of encouragement to push me past my mental and physical boundaries.

Beyond the academic and clinical aspects, I discovered a newfound sense of self-reliance. The demanding curriculum demanded meticulous time management, effective study habits, and a remarkable ability to prioritize tasks. Learning to balance my academic workload with my personal life required a level of self-discipline and organizational skills I had never thought possible. I learned to prioritize effectively, and to cultivate a healthy balance between intense study and essential self-care. This self-discipline extended beyond the walls of the veterinary school. I realized that success was less about sheer intelligence and more about persistent effort and resourceful strategizing.

The rigorous curriculum wasn't just about acquiring knowledge; it was about cultivating a mindset of continuous learning and critical thinking. Veterinary medicine is a dynamic field that constantly develops with new research and technological advancements. To succeed, one needs to be adaptable, willing to embrace change, and committed to lifelong learning. Veterinary school instilled in me this mindset, preparing me not only for the challenges of my career but also for the lifelong journey of professional development. It was a crucible, burning away superficial understanding and leaving behind a deeper appreciation for the complexities of life, both in animals and humans—the emotional challenges, intellectual rigor, and demanding practical work instilled in me an unwavering resilience.

By the end of my first year, I had undergone a profound transformation. A quiet confidence, a deep sense of accomplishment, and an unwavering belief in my ability to succeed had replaced the initial overwhelming sense of anxiety and self-doubt. The journey had been demanding, and emotionally challenging, but it had also been gratifying. It had pushed me to my limits, revealing hidden strengths and fostering a level of resilience I hadn't known I possessed. I had not only survived the rigors of veterinary school, but I had thrived. I conquered the challenging landscape of anatomy and physiology, not through brute force, but through strategic planning, perseverance, and an unyielding belief in my capacity to learn and adapt. With the foundation established, I awaited the next academic challenge. The exhaustion was still present, the sleep deprivation a constant companion, but a triumphant pride now accompanied it. This was just the beginning of my journey; the next chapter was yet to come.

Chapter 12: Clinical Rotations: Real World Experience

The transition from the theoretical world of textbooks and cadaver labs to the vibrant chaos of clinical rotations was both jarring and exhilarating. My first rotation was at a bustling equine practice nestled in the rolling hills outside of Ithaca. The scale was immediately different. Instead of the meticulously controlled environment of the university hospital, the unpredictable world of large animals thrust me into its chaos; temperamental horses, erratic weather, and demanding farm schedules were the norm. My initial apprehension quickly morphed into fascination as I learned the intricacies of equine medicine, from lameness examinations and dental procedures to the complexities of colic surgery and reproductive management.

I remember my first solo exam of a majestic chestnut mare suspected of founder, a debilitating hoof condition. My hands trembled as I palpated her legs for heat, swelling, and pain. The mare, sensing my nervousness, shifted uneasily. Under the calm, watchful guidance of Dr. Ramirez, I systematically documented my findings. Sadly, the diagnosis confirmed founder. The subsequent treatment, which involved meticulous hoof trimming and pain management, required a delicate balance of expertise and empathy. Seeing the relief on the owner's face when the mare showed signs of improvement was deeply gratifying. This experience solidified my commitment to my chosen career path and made the long hours of study and hard work feel worthwhile.

My next rotation was a stark contrast–a small animal practice in a nearby town. The pace was frenetic, a whirlwind of vaccinations, routine check-ups, and emergency cases. Here, I encountered the full spectrum of small animal medicine, from treating a playful kitten with a minor upper respiratory infection to managing a critical case of canine pancreatitis. The emotional toll was intense. One case that particularly stands out involved a senior golden retriever named Gus, suffering from a rapidly progressing lymphoma. Watching his gradual decline, the gradual dimming of his playful spirit, was emotionally draining. Communicating the diagnosis and the prognosis to his heartbroken owners was one of the most challenging experiences of my clinical rotations. There was a palpable sense of grief in the room. It was a harsh lesson in the realities of veterinary medicine: the bittersweet mix of healing and loss, of victories and heartbreaking farewells.

The sheer variety of cases I encountered during this rotation broadened my practical skills and expanded my understanding of veterinary medicine. My abilities in animal venipuncture, medication administration, and diagnostic test interpretation have significantly improved. My skills in radiograph interpretation advanced to where I could precisely and accurately identify subtle fractures, foreign bodies, and other abnormalities. Assisting in minor surgeries, including spays and neuters, I developed my surgical skills, particularly in tissue handling and suture placement. My confidence soared with each successful procedure, producing a feeling of accomplishment surpassing academic achievements. The practice's holistic approach–incorporating alternative therapies and behavioral modifications alongside traditional veterinary medicine–particularly impressed me. The way acupuncture, physical therapy, and herbal remedies complemented conventional treatments, particularly in managing chronic conditions like arthritis, fascinated me. This experience broadened my understanding of the multifaceted nature of animal care.

My surgical rotation at the university teaching hospital was a high-pressure environment characterized by a rapid-fire sequence of complex cases. The atmosphere was a vibrant blend of urgency and focused determination, a whirlwind of beeping monitors, hushed conversations, and the rhythmic whir of surgical equipment. The wide range of surgical techniques, from routine orthopedic procedures to complex cardiothoracic surgeries, exposed me to them.

I learned the importance of meticulous preparation, precise technique, and teamwork in a high-stakes surgical setting. It was here that I witnessed firsthand the seamless integration of clinical skills and academic knowledge, as well as the crucial role of a veterinary team. The senior surgeons, with their years of accumulated experience and expertise, served as extraordinary mentors, guiding me through the intricacies of each case, highlighting the subtle nuances of surgical technique, and emphasizing the critical importance of pre-operative planning and post-operative care. Each case was a learning experience, a window into the world of complex surgical interventions and the crucial role of collaboration, patience, and precision.

Equally valuable, though vastly different, was my experience at a wildlife rehabilitation center. Working with injured and orphaned wild animals required a unique set of skills, combining expertise in wildlife biology with a deep understanding of animal behavior. The emphasis shifted from the conventional treatments found in traditional veterinary practices to a more naturalistic approach, emphasizing patient rehabilitation and release back into their natural habitat. I learned to handle a diverse range of species, including raptors, mammals, and reptiles, developing a deep respect for their unique vulnerabilities and behavioral complexities. This experience shaped my understanding of conservation medicine, highlighting the intersection of animal health and environmental stewardship.

Every clinical rotation presented unique challenges, but they all contributed significantly to my growth as a veterinary student. The late nights, the demanding workloads, and the emotional toll–these were all part of the process. What emerged from these experiences was not merely an accumulation of technical skills, but a profound understanding of the complex relationship between animals, their owners, and the veterinary profession. Navigating the emotional landscape of veterinary medicine, I learned to balance empathy with professionalism and find strength in vulnerability. Through experience, I have developed the ability to communicate effectively, conveying both hope and realism with sensitivity and respect, to owners who are distraught.

My training has equipped me with the skills to navigate dynamic situations, exercising critical thinking under pressure to make timely, informed decisions despite incomplete information. Teamwork's importance, the interconnectedness of roles in a veterinary hospital, and the cumulative expertise of a dedicated team became clear to me. Far exceeding mere periods of practical experience, these rotations served as crucial platforms for professional development and the acquisition of practical skills. They were an affirmation of my vocation–not merely a career choice, but a calling. The journey was far from over, yet the experiences gained during these rotations had solidified my conviction that this path was the one for me. The challenges had been significant, the emotional toll considerable, but it had been an enriching and profoundly transformative experience. I was ready for the next stage of my journey–the culmination of years of hard work and dedication.

Chapter 13: Building Relationships, Working with Colleagues and Mentors

The final year of veterinary school felt like a sprint to the finish line, a whirlwind of advanced coursework, increasingly complex clinical rotations, and the looming shadow of board exams. Yet, amidst the academic pressure and the relentless pace, I discovered the profound importance of building strong professional relationships. It wasn't just about mastering the intricacies of veterinary medicine; it was about learning to navigate the profession's complexities through collaboration and mentorship.

My relationship with Dr. Anya Sharma, a renowned surgeon specializing in small animal orthopedics, exemplifies the power of mentorship. Dr. Sharma wasn't just a brilliant surgeon; she was a patient and insightful teacher who understood the balance between pushing her students and fostering their growth. During my surgical rotation under her tutelage, she took the time to explain the rationale behind every surgical decision, patiently answering my questions. She encouraged me to think critically, to question established protocols, and to develop my surgical intuition.

I vividly recall a challenging case involving a dachshund with a complex fracture. Dr. Sharma guided me through the pre-operative planning, allowing me to take part in the decision-making process, from selecting the appropriate surgical approach to choosing the correct implants. The surgery itself was intricate, requiring precision and unwavering focus. Dr. Sharma's calm presence and steadfast support were invaluable during the procedure, and her post-operative instructions were meticulous. The successful outcome of the surgery was a testament not only to Dr. Sharma's expertise but also to the supportive learning environment she fostered. Her mentorship extended beyond the operating room. She invited me to attend her research meetings, exposing me to the world of veterinary research and the scientific inquiry process. These interactions instilled in me a passion for research, shaping my future academic pursuits.

Her guidance helped me to approach surgery not only with technical skill but also with empathy and compassion, understanding the emotional weight the procedure held for both the animal and its owner.

Beyond individual mentorship, the importance of teamwork became increasingly apparent. The collaborative nature of veterinary medicine was evident in every aspect of my clinical rotations. During my emergency medicine rotation, I learned firsthand the value of a well-coordinated team. The fast-paced environment demanded seamless collaboration between veterinarians, veterinary technicians, and support staff. It was a dynamic orchestra of skill and expertise, where everyone played a critical role in providing timely and effective care. I remember a chaotic night; a car hit a dog. The entire team sprang into action.

The veterinary technicians efficiently prepared the patient, while the veterinarian assessed the situation and organized the initial diagnostic tests. I worked alongside another veterinary student, coordinating the administration of fluids and medications under the watchful eye of our supervising veterinarian. Communication was key, with concise updates exchanged between team members, ensuring everyone agreed. The collective expertise of the team, fueled by their collaborative spirit, successfully stabilized the dog, giving him a chance to survive. This experience solidified my appreciation for the power of teamwork, underscoring the interconnectedness of roles and the collaboration that emerges from a shared purpose and effective communication.

This emphasis on teamwork extended beyond the immediate clinical settings. I developed invaluable friendships with my classmates, forming a tight-knit support network. Late-night study sessions became a ritual, fueled by copious amounts of coffee and mutual encouragement. We quizzed each other, shared notes, and offered emotional support during particularly challenging times.

These relationships were critical in navigating the stresses of veterinary school. Shared burdens, collective struggles, and shared victories cemented the bonds and strengthened them. Our shared experiences provided invaluable strength, helping us overcome academic pressure, personal setbacks, and the emotional toll of clinical rotations. The bond we forged transcended the educational realm; it was a testament to the importance of community and camaraderie in a demanding and emotionally challenging profession. We shared not only academic knowledge but also our vulnerabilities, our joys, and our fears. We celebrated each other's successes and offered solace during moments of doubt and disappointment. These friendships created a safety net—a supportive community that not only helped us survive but also thrive.

My interactions with professors extended beyond the lecture hall. I actively sought faculty members whose research interests aligned with my own, attending their office hours not just for academic help but also to engage in intellectual discussions. Professor David Miller, an expert in wildlife conservation medicine, opened doors to opportunities I hadn't even imagined. His guidance helped me secure a research position in his lab, allowing me to contribute to a project focused on the impact of climate change on wildlife populations. This experience proved invaluable in refining my research skills and broadening my understanding of the role of veterinary medicine in conservation. Beyond specific projects, Professor Miller instilled in me a deeper appreciation of the broader ethical and societal implications of veterinary science. He supported the importance of advocacy, emphasizing the veterinarians' responsibility to be active stewards of animal welfare.

The relationships I built throughout veterinary school extended beyond the professional realm, enriching my life beyond my academic pursuits. The support system I developed–my mentors, colleagues, friends, and professors–wasn't just about academic success; it was about navigating the emotional complexities of a challenging and demanding profession. These individuals provided an invaluable source of support, guidance, and encouragement. It was the combination of their support and my relentless pursuit that allowed me to navigate the challenges and achieve my goal. The journey to becoming a veterinarian was not a solo endeavor; it was a collaborative effort, a testament to the power of mentorship, teamwork, and the unwavering support of a dedicated community. The relationships forged during this period would prove invaluable, shaping not only my professional life but also my personal growth and sense of purpose. They were not just professional connections, but meaningful relationships that enriched my life in ways I never expected. The culmination of these experiences laid a solid foundation for my future career as a veterinarian.

Chapter 14: Research Opportunities
Scientific Contributions

The final year of veterinary school culminated in a flurry of activity–board exams loomed, clinical rotations intensified, and the sheer volume of information felt overwhelming. Yet, amidst the pressure, a new avenue of exploration opened up in research. That space allowed me to apply and expand upon the theoretical knowledge I'd absorbed. My involvement in research wasn't simply about adding another line to my CV; it was a passionate pursuit that sparked a deep-seated intellectual curiosity.

My initial foray into research began unexpectedly. Professor Miller, whose lectures on wildlife conservation had captivated me, invited me to join his team working on a project studying the impact of deforestation on the migratory patterns of Canada geese. It was a far cry from the controlled environment of the surgical suite or the fast-paced urgency of the emergency room. Instead, it involved long days in the field, collecting data in unpredictable weather, and dealing with the inherent challenges of working with wild animals.

The project involved meticulous data collection, requiring precise observation and recording of goose behavior, habitat usage, and migratory routes. We used GPS tracking devices attached to a small sample of geese to monitor their movements over several months. This involved careful planning, coordination with park authorities, and countless hours of painstaking data analysis. The raw data comprised thousands of GPS coordinates, which we then processed using specialized software to create detailed maps of their migratory paths. This stage involved learning new software, mastering statistical analysis techniques, and collaborating closely with other members of the research team. The team comprised not just veterinary professionals but also biologists, geographers, and statisticians, each bringing their unique expertise to the project.

This interdisciplinary approach highlighted the complexity of environmental research and underscored the need for collaborative efforts to comprehend the multifaceted effects of human activities on wildlife.

Analyzing the data revealed some fascinating—and concerning—results. We found that geese deforestation exhibited altered behavior, characterized by longer migration times, significant disruption of their traditional migratory routes, increased energy expenditure, and a higher incidence of stress indicators. This underscored the critical link between habitat loss and animal well-being. My contribution to the project involved developing a predictive model to forecast the potential impact of future deforestation scenarios on the geese's migratory patterns. This involved applying complex statistical modeling techniques, refining my understanding of population dynamics, and enhancing my programming skills. Using the model, we simulated various scenarios and projected potential consequences, gaining valuable insights for conservation efforts.

The experience wasn't without its setbacks. Several GPS trackers failed, forcing us to adapt our data collection methods. There were days when inclement weather hindered fieldwork, causing delays in progress. There were also challenges related to data analysis, requiring extensive troubleshooting and revision of the predictive model. These hurdles weren't just frustrating; they were valuable learning experiences, teaching me the importance of adaptability, resilience, and meticulous attention to detail. Presenting our findings at a national veterinary conference was an enriching experience. The research generated significant interest, leading to further collaborations and grant applications. The study's impact extended beyond the academic realm; its results informed regional conservation policies, highlighting the tangible effect of research on real-world problems.

Beyond the Canada goose project, I also became involved in a separate research initiative focused on developing novel antibiotics to combat antibiotic-resistant bacteria in companion animals. This project, led by Dr. Emily Carter, a microbiologist in the veterinary school, involved laboratory work, meticulous experimentation, and extensive data analysis. The work was demanding and required specialized training, but it ignited a passion for microbiology and infectious disease control. Our research focused on a specific strain of bacteria known for its resistance to multiple antibiotics, a growing concern in veterinary medicine. My role involved testing the effectiveness of the newly

We synthesized antibiotics on this resistant strain using advanced microbiological techniques. The process demanded precise measurements, sterile procedures, and careful data recording. We tested compounds at multiple concentrations, monitoring their effects on bacterial growth and viability, performing numerous assays, and analyzing the results statistically. The project honed my lab skills, deepened my understanding of infectious diseases, and sparked a lasting appreciation for the complexities of antibiotic resistance.

The challenges in this research project were many. One of the most frustrating obstacles was the variability in experimental results, which required meticulous control of variables and repeated testing to ensure accuracy. The process was time-consuming, often requiring late nights in the lab, but the potential benefits fueled our perseverance. Several compounds showed promising results, demonstrating effective inhibition of bacterial growth even in the presence of antibiotic resistance mechanisms. Our presentation of these findings at an international veterinary microbiology conference generated significant interest from researchers and pharmaceutical companies. The successful outcome underscored the importance of ongoing research in developing novel antibiotics to combat the escalating threat of antimicrobial resistance. The collaboration with Dr. Carter provided invaluable mentorship that extended beyond the laboratory setting. She encouraged independent thinking, nurtured my scientific curiosity, and provided guidance in developing my research skills.

The experience gained through these research projects extended far beyond the acquisition of technical skills. I learned critical evaluation of scientific literature, well-controlled experiment design, and complex dataset analysis. I also learned the importance of teamwork, collaboration, and clear communication in a scientific setting. These experiences not only shaped my professional identity but also influenced my approach to problem-solving and my understanding of the power of scientific inquiry.

My involvement in research extended beyond laboratory work and fieldwork to include scientific writing and presentation. Research reports, grant applications, and conference presentations improved communication and the effective conveyance of complex science to diverse audiences. The meticulous preparation required for scientific presentations, the careful structuring of research reports, and the need for clear and concise communication honed my ability to synthesize and communicate information effectively, skills vital not just for my future research endeavors but for my overall career as a veterinarian.

The culmination of these research experiences, which broadened my scientific knowledge and refined my technical skills, significantly affected my career trajectory. Through my research, publications strengthened my graduate school applications, while gaining research skills became invaluable assets for my doctorate. The research not only contributed scientifically to the field of veterinary medicine but also represented a personal triumph—a testament to my dedication, perseverance, and the boundless support of my mentors and colleagues.

Chapter 15: Overcoming Challenges: Time Management and Self-Care

The relentless pace of veterinary school demanded more than just intellectual prowess; it required mastery of time management and a deep commitment to self-care. The sheer volume of coursework, practical sessions, and extracurricular activities threatened to overwhelm me — a constant pressure cooker simmering just below the boiling point. Chaotic schedules, fueled by caffeine and a frantic attempt to juggle everything, characterized my early days. Sleep deprivation became the norm, driven by the comforting illusion that pulling all-nighters was a badge of honor, proof of unwavering dedication. My apartment, once a haven of tranquility, transformed into a chaotic study zone littered with textbooks, notes scribbled on every available surface, and half-eaten containers of takeout. This unsustainable lifestyle eventually took its toll, manifesting as exhaustion, irritability, and a general sense of impending doom.

My breaking point arrived during a fierce week. The pressure of three major exams, a difficult clinical rotation, and an upcoming research paper deadline combined to create an overwhelming amount of stress. I felt utterly overwhelmed; my body ached with fatigue, and my mind was a swirling vortex of anxieties and responsibilities. I missed classes, neglected my friendships, and even struggled to maintain basic hygiene. The realization that I was spiraling out of control hit me with the force of a physical blow. I knew something had to change; I needed to find a better way to navigate this relentless academic marathon.

The first step was acknowledging the problem and accepting that I couldn't—and shouldn't—try to do everything at once. The notion that constant striving was a sign of success had to be unlearned. I started by creating a realistic schedule, prioritizing tasks based on urgency and importance. This meant learning to say "no" to some commitments, a skill I had desperately needed to develop. A sense of relief gradually replaced the guilt I initially felt about relinquishing and improved focus. I started using a planner not just to track assignments and deadlines but also to schedule time for meals, exercise, and relaxation. This deliberate act of self-care initially felt like an indulgence, a luxury I couldn't afford, but it quickly became a necessity.

Creating a realistic schedule was only the beginning. I experimented with various time management techniques, eventually settling on a combination of the Pomodoro Technique and time blocking. The Pomodoro Technique, which involves focused work intervals interspersed with temporary breaks, helped me maintain concentration and prevent burnout. Time blocking, allocating specific time slots for particular tasks, provided a framework for my day, reducing the feeling of being constantly overwhelmed by a never-ending to-do list. I also discovered the power of prioritizing tasks; it wasn't enough to list them—I needed to categorize them according to their importance and urgency. Focusing on the most critical tasks first allowed me to complete the most crucial things, even if it meant postponing less essential things.

Simultaneously, I embarked on a journey of self-discovery, recognizing that my well-being wasn't a separate entity from my academic pursuits; instead, it was an integral part of them. Ignoring my physical and mental health had been a detrimental approach, akin to trying to drive a car with a flat tire. I started incorporating regular exercise into my schedule, initially starting with short walks to clear my head and escalating the intensity and duration. Exercise became not just a physical activity, but a mental escape, a time to clear my mind and de-stress. I discovered the joy of yoga, finding solace in the mindfulness and physical movement. The practice provided a much-needed counterpoint to the intense mental exertion of my studies. It was a space where I could release tension, reconnect with my body, and cultivate a sense of inner peace.

Nutrition also underwent a significant transformation. My diet, previously dominated by caffeine and convenience food, needed a complete overhaul. I started focusing on whole, unprocessed foods, ensuring that I was fueling my body with the nutrients it needed to cope with the demands of veterinary school. This involved meal prepping on the weekends, planning nutritious meals to avoid impulsive, unhealthy choices during the hectic week. Hydration became a priority, with a focus on replacing sugary drinks with water and herbal teas. The change was gradual, but the difference in my energy levels and overall well-being was remarkable.

Beyond physical and nutritional changes, I understood that mental well-being was equally crucial. I started practicing mindfulness meditation, even if it was only for a few minutes each day. The practice of focusing on the present moment, without judgment, helped reduce my anxiety and improve my ability to cope with stress. I also found solace in creative pursuits, such as sketching and painting, which provided a welcome escape from the academic pressure cooker. They allowed me to express myself creatively, fostering a sense of calm and emotional balance.

Sleep became a non-negotiable priority. I gradually shifted my sleeping pattern, aiming for seven to eight hours of uninterrupted sleep each night. This involved establishing a consistent bedtime routine, creating a relaxing sleep environment, and limiting screen time before bed. The impact was profound; I experienced improved concentration, reduced stress, and heightened emotional resilience.

Social connections also played a pivotal role in maintaining my well-being. My friendships, initially neglected during the most intense periods, became a source of strength and support. Spending time with friends and engaging in activities outside of my studies helped maintain my perspective and a sense of normalcy. I learned the importance of open communication and the value of seeking support from friends, family, and mentors when needed. This involved actively expressing my struggles and seeking advice from those who understood the pressures of veterinary school.

One couldn't overstate the importance of setting boundaries.

Learning to say "no" to additional commitments, even if they seemed appealing or beneficial, became essential. The ability to discern between valuable obligations and those that added to the already overwhelming workload was a skill I actively cultivated. Saying "no" wasn't an act of selfishness; it was an act of self-preservation, ensuring that I prioritized the most crucial tasks and maintains a sustainable pace.

One of the most valuable lessons I learned was the importance of self-compassion. I acknowledged I was not a superhuman entity capable of effortlessly navigating every challenge. I allowed myself to make mistakes, to experience setbacks, and to feel overwhelmed. Rather than berating myself for imperfections, I chose self-forgiveness, acknowledging that setbacks were inevitable and that it was okay to falter along the way.

Through these strategic time management techniques and deliberate self-care practices, I transformed my approach to veterinary school. The initial chaotic whirlwind gradually gave way to a more balanced and sustainable existence. I discovered that prioritizing well-being showed self-awareness and proved that academic success shouldn't compromise my mental and physical health. This carefully constructed balance proved to be the cornerstone of my success, allowing me to thrive academically while simultaneously nurturing my overall well-being. It was a significant fundamental change, one that transformed a daunting challenge into an ultimately rewarding and fulfilling journey. This newfound equilibrium wasn't simply about managing time and stress; it was about cultivating a sustainable lifestyle that allowed me to flourish, not just as a student but as a person. This holistic approach became my guiding principle, influencing not only my academic pursuits but every facet of my life thereafter.

Chapter 16: Choosing a Specialization: Focusing on the Path

As the final year of veterinary school began, it arrived much like a seasoned marathon runner, completely exhausted from prior years but resolute in its determination to finish the race. The frenetic pace was feeling less like a chaotic sprint and more like a sustained, though strenuous, jog. As the years of intense study neared their end, I felt both thrilled and apprehensive. Graduation thrilled me, yet the uncertainty of the future weighed heavily on me. A doctorate seemed the logical next step, yet the road ahead appeared as daunting and intricate as the veterinary field. The sheer number of specialization choices paralyzed me. Choosing a career wasn't the only thing involved;

My early fascination with veterinary science stemmed from a deep-seated love for animals, a childhood passion nurtured by countless hours spent observing the surrounding creatures–the playful antics of neighborhood dogs, the graceful flight of birds, the silent grace of a wild deer. This wasn't just a simple fondness, but a genuine, deep-rooted connection. This foundation solidified during my participation in veterinary science competitions, fostering a competitive spirit while also deepening my understanding of animal health and welfare. During this time, I planted the seed of a professional veterinary career, nurturing it through intense study and hands-on practice. Now, on the precipice of a doctorate, I have refined my passion through years of dedicated learning and real-world experience.

The process of choosing a specialization felt like standing at the edge of an immense forest, each path promising a fresh adventure, a unique journey of discovery. Cardiology, with its intricate mysteries of the heart, beckoned with a particular allure. The delicate balance of the circulatory system, the subtle intricacies of heart function, and the challenge of diagnosing and treating life-threatening cardiac conditions presented an intellectually stimulating pathway. Yet, my fascination extended beyond the realm of purely technical medical prowess; it lay in the delicate interplay between physical diagnosis, technological advancements, and the emotional connection with the animal patient and its owner.

Equine veterinary medicine held a similar appeal. The majestic power of horses, their unique physiology, and the intricate challenges of caring for such magnificent creatures deeply resonated with me. The bond between horse and rider, the athleticism of these animals, along with the unique challenges of managing their health within the context of their athletic endeavors, all contributed to their appeal. I imagined myself on a sprawling ranch, amidst the beauty of open fields, skillfully assessing the health of magnificent horses, providing treatment, and working alongside dedicated horsemen and horsewomen.

Minor animal surgery, with its meticulous techniques and the profound satisfaction of restoring an animal's health through surgical intervention, also captured my attention. The challenge of intricate procedures, the technical proficiency demanded by this specialization, and the knowledge that my work would directly affect the animals' well-being served as a powerful draw. The palpable sense of fulfillment witnessed in the post-operative recovery of animals under my care was something I longed to experience actively.

However, it was the field of wildlife veterinary medicine that ultimately captured my imagination and my heart. The raw, untamed beauty of the wild, the unique challenges of treating animals in their natural habitats, and the vast unknowns that the field presented–all deeply captivated me. This specialization didn't just involve treating sick animals; it encompassed a broader dedication to conservation, to the preservation of delicate ecosystems, and to protecting vulnerable species. The blend of clinical expertise, scientific research, and conservation advocacy perfectly aligned with my values and long-term aspirations. The prospect of working in remote locations, immersing myself in nature, and playing a direct role in protecting wild animals proved irresistible.

I arrived at my decision after a period of considerable contemplation. The initial step on my professional journey was the necessity of successfully completing the state veterinary boards. Because I dedicated myself to my studies, I was successful in obtaining my state license as a Doctor of Veterinary

Medicine. I found myself when I had to make a hard decision. Over a period of several months, I investigated different veterinary specializations, which involved conducting research, attending conferences, taking part in interviews with professionals in the field, and gaining hands-on experience by shadowing them. In order to determine which veterinary specialty would best use my skills, I comprehensively reviewed current scientific publications. I spent some time considering my interests, my personality traits, and my aspirations for a future career. Given the contrasting environments of a private clinic's structure and the often unpredictable nature of fieldwork, which one aligned better with the needs and preferences? The question remained whether my preference was more aligned with intricate surgical procedures, or did veterinary public health held a stronger appeal for me?

The process involved intense self-reflection. I scrutinized my strengths and weaknesses, assessing my level of comfort with different aspects of veterinary medicine. Did I possess the technical proficiency for surgery? Was I adaptable enough to handle the unpredictable nature of fieldwork in remote locations? I carefully examined the specific requirements of each specialization, considering the coursework, the practical experience needed, and the potential career pathways each offered.

I also considered the lifestyle implications. Would I be willing to work long, irregular hours, often in challenging environments? Did I have the fortitude to work in isolated areas, often far from family and friends? The balance between professional fulfillment and personal well-being needed careful consideration.

Ultimately, a confluence of factors influenced my decision to pursue a doctorate in wildlife veterinary medicine. It was the culmination of my childhood passion for animals, my fascination with the intricate workings of the animal body, my academic achievements and practical skills, as well as my deep-seated commitment to conservation and environmental protection. The challenge of navigating this field, both academically and professionally, excited me. It was more than a career choice; it represented a profound commitment to a life's work that combined scientific rigor with a genuine, unwavering love for the natural world.

Choosing this path didn't ease the daunting task ahead. The path to a doctorate in wildlife veterinary medicine wouldn't be easy. It would require a profound understanding of various animal species and their behaviors, rigorous training in specialized veterinary techniques adaptable to diverse environments, extensive fieldwork experience, and an unwavering commitment to advancing our understanding of animal health within the context of ecosystem health. I knew I would face immense challenges, long hours in demanding environments, and the emotional weight of treating animals in often precarious situations. But these challenges, rather than deterring me, fueled my resolve. The potential to contribute meaningfully to wildlife conservation, protect endangered species, and play a role in preserving the natural world made the daunting task worthwhile.

Completing the application felt like an adventure. I carefully prepared my materials, crafting a statement of purpose that showcased my enthusiasm for wildlife veterinary medicine while highlighting my academic achievements and relevant experiences. Professors, researchers, and mentors who know my qualifications and commitment wrote letters of recommendation for me.

The wait for acceptance was agonizing — a period of intense anticipation that tested my patience and resilience. Each day felt like an eternity, as I anxiously awaited news that would determine the next chapter of my life. When the acceptance letter finally arrived, the joy was overwhelming — a validation of years of hard work, dedication, and unwavering perseverance. The feeling was like the exhilaration of a hard-fought victory in a veterinary science competition–a triumph of preparation, skill, and determination. However, this victory marked not just an end but a significant beginning–the start of a new phase in my journey, one that promised to be equally demanding, challenging, and ultimately profoundly rewarding.

The formal commencement of my doctoral program marked a significant transition in my life. It wasn't just a change of environment; it was a metamorphosis of my personal and professional identity. Challenging coursework pushed my knowledge to its limits, expanding my understanding of the intricate world of wildlife medicine. The research component involved a thorough examination of the scientific literature, with many hours spent planning hypotheses, designing research protocols, and meticulously analyzing the data. The fieldwork was equally demanding. I found myself in remote locations, often battling challenging terrain, extreme weather, and the uncertainties of wildlife encounters. But these challenges became growth opportunities, moments that tested my adaptability, resilience, and dedication to my chosen path.

The mentorship I received was invaluable, providing guidance, support, and encouragement throughout the program. My professors, researchers, and senior colleagues offered perspectives that reshaped the way I saw things, shared their expertise, and helped me navigate the complexities of both my academic and professional journeys. They became not just teachers but also mentors, guiding my progress, offering advice, and shaping my understanding of the field. Their guidance provided a compass, guiding me through the often-uncertain path toward achieving my professional goals.

I learned the importance of patience, perseverance, and the unwavering pursuit of knowledge. There were moments of frustration, moments of self-doubt, and times when the sheer enormity of the task ahead seemed overwhelming. However, amidst the challenges, I found strength in my passion, in the guidance of my mentors, and in the unwavering support of my family and friends. Their belief in my capabilities served as a constant source of encouragement, helping me overcome obstacles and maintain my focus on the goal. It was this intricate tapestry of hard work, determination, and steadfast support that enabled me to navigate the complexities of my doctoral journey and emerge stronger, more resilient, and more deeply committed to my chosen field. The road ahead was still long, but I was ready. The journey of becoming a wildlife veterinarian had only just begun.

Chapter 17: Doctoral Research: In-depth Investigation

The thrill of acceptance quickly gave way to the sobering reality of doctoral research. My chosen topic — the impact of climate change on the migratory patterns of the Arctic fox — felt both exhilarating and terrifying in equal measure. The sheer scope of the project was daunting. I knew I wouldn't confine my research to a sterile laboratory environment.

The initial stages of the research involved an exhaustive literature review, as well as a comprehensive examination of existing studies on Arctic foxes, climate change, and their interconnection. I spent months immersed in scientific journals, analyzing data, and familiarizing myself with established methodologies. The sheer volume of information was overwhelming — a vast ocean of data that threatened to swallow me whole. I learned to navigate this ocean strategically, focusing on key publications and refining my research question. My advisor, Dr. Anya Sharma, a renowned expert in wildlife conservation, provided invaluable guidance during this phase. Her insights and suggestions were critical in shaping my research design and refining my focus. Her mentorship was more than just academic guidance; it was a constant source of encouragement and support during moments of doubt and frustration.

Developing a robust research method proved to be a significant challenge. The Arctic environment is unforgiving, with unpredictable weather patterns, extreme temperatures, and limited access to resources. My research design needed to account for these logistical hurdles. I had to develop a data collection strategy that was both workable and reliable under challenging circumstances.

This involved deliberating the timing of my fieldwork expeditions, the technology to be employed, and the means of ensuring the safety and well-being of both myself and the research team. The process involved meticulous planning, collaboration with experienced field researchers, and a deep understanding of the Arctic ecosystem. I spent countless hours designing data collection protocols, testing equipment, and perfecting my approach to data analysis. Each step involved rigorous evaluation and refinement, ensuring accuracy and reliability despite environmental constraints.

The fieldwork itself was an incredible journey, albeit a physically and mentally taxing one. I spent months in the Arctic, enduring blizzards, freezing temperatures, and the constant threat of unpredictable weather. The landscape was breathtaking, a stunning panorama of snow-covered mountains, icy plains, and frozen rivers.

However, the beauty of the environment belied its unforgiving nature. Treacherous terrain demanded long days of trekking, often in sub-zero temperatures. I carried heavy backpacks filled with research equipment, enduring physical discomfort and fatigue. The isolation was challenging, a stark reminder of my distance from family, friends, and the comforts of civilization.

Data collection involved tracking the movements of Arctic foxes using GPS collars, collecting fecal samples for genetic analysis, and monitoring their hunting behavior. This required patience, perseverance, and a deep understanding of the animals' behavior. The Arctic foxes were elusive; their movements were unpredictable. I often spent hours observing them from a distance, meticulously documenting their activities, and carefully collecting samples without disturbing their natural behavior. The process involved painstaking attention to detail and a constant awareness of the delicate balance of the Arctic ecosystem.

The challenges I encountered during fieldwork were many. Equipment malfunctions, unexpected weather events, and the logistical difficulties of working in a remote environment were constant sources of frustration. Sometimes I questioned my ability to complete the research. However, my determination, the support of my research team, and the guidance of my mentor saw me through these challenging moments. My teammates, a diverse group of researchers from around the world, became like family to me during the months we spent in the Arctic. Their support, camaraderie, and unwavering belief in the project proved invaluable, helping me overcome setbacks and maintain my focus on the research goals.

Analyzing the data was equally complex. I applied advanced statistical techniques to uncover trends in Arctic fox migration and correlate them with climate change indicators. The process involved extensive modeling, rigorous analysis, and countless hours interpreting results. Though frustrating at times, with perseverance and guidance from my statistical advisor, I was able to extract meaningful insights. My findings revealed a significant correlation between rising temperatures and changes in the Arctic foxes' migratory behavior, showing the potential vulnerability of these animals to the effects of climate change. This highlighted the urgent need for conservation efforts to protect this vulnerable species.

The writing of my dissertation was the last hurdle. I spent months crafting an interesting narrative, integrating my findings into a coherent whole. The process was demanding, requiring meticulous attention to detail, clear articulation of complex scientific concepts, and careful consideration of the broader implications of my research. I revised and edited my work many times, continually striving to enhance its clarity, precision, and overall impact. The completion of my dissertation was not simply the end of my doctoral research; it marked a culmination of years of hard work, dedication, and perseverance. The feeling was immense, a blend of relief, satisfaction, and profound gratitude for the support I had received along the way.

Defending my dissertation was both challenging and rewarding. I presented my research to a panel of esteemed experts, who subjected my work to rigorous scrutiny. The questions were hard, pushing me to the boundaries of my knowledge. However, I was well-prepared. The thorough research process, the countless hours spent studying, and the feedback from my mentors had prepared me for this moment. I answered each question with clarity and confidence, defending my method and interpreting my findings.

Successfully defending my dissertation was more than an academic achievement; it marked a pivotal milestone in my personal and professional journey. It reflected my resilience, dedication to research, and the value of mentorship and collaboration. This moment validated the countless hours of study, unwavering commitment, and deep passion that had driven me toward becoming a wildlife veterinarian. The Arctic fox research, far from being a mere project, had become a pivotal point in my life and career, solidifying my commitment to wildlife conservation and laying the groundwork for my next endeavors.

Chapter 18: Conference Presentations and Publications: Sharing Knowledge

The exhilaration of completing my dissertation was short-lived, quickly replaced by the anticipation of sharing my findings with the broader scientific community. My advisor, Dr. Sharma, had already encouraged me to submit an abstract to the annual Wildlife Conservation Society conference in Banff. This prestigious event attracts leading researchers from around the globe. The thought was both thrilling and terrifying. Presenting my research to such a distinguished audience was a daunting prospect—a leap of faith into the critical eye of my peers.

Putting together the conference presentation proved to be quite demanding. Months of research went into creating my concise and engaging slides, which I meticulously crafted. To maximize impact and clarity, I painstakingly selected each graph, chart, and image. To ensure a smooth and confident delivery, I practiced my presentation extensively, refining my approach. Summarizing the extensive data was difficult, requiring me to focus on key results and simplify my descriptions. My goal was to communicate my research's importance, the impact of my findings, and the urgency of conservation in just twenty minutes.

The day of the presentation arrived with a mix of excitement and nerves. The conference hall buzzed with activity, creating a vibrant tapestry of scientific minds engaged in lively discussions. I felt a knot of anxiety in my stomach, a familiar feeling that was amplified by the weight of the occasion. As I waited backstage, I reviewed my notes one last time, reminding myself of the countless hours I had invested in this research. My heart pounded in my chest as my name was called, signaling my entry onto the stage.

The spotlight felt intensely bright, the audience a sea of faces blurred in the dim light. I began my presentation, my voice initially trembling slightly, but quickly gaining strength and confidence as I delved into my research. I presented my findings clearly and concisely, supported by interesting visuals and data. The audience was engaged, their attention captivated by the significance of my work.

I could feel their interest in the dramatic changes witnessed in the Arctic foxes' migratory patterns and the clear implications of climate change. The Q&A session was both challenging and rewarding. The questions from the audience were insightful and probing, forcing me to think critically about my research and its limitations. I answered each question thoughtfully and thoroughly, defending my method and interpretations. The experience was invaluable, providing an opportunity to engage with leading experts in my field, receive constructive criticism, and refine my understanding of the research.

The feedback was overwhelmingly positive. Attendees praised the thoroughness of my research, the clarity of my presentation, and the significance of my findings. Several senior researchers approached me afterward to discuss future collaborations. The conference not only showcased my work but also offered valuable opportunities for networking and building professional connections.

Following the success of my conference presentation, I began working on a manuscript for publication in a peer-reviewed journal. The journal selection process involved careful consideration; I chose a journal known for its high impact factor and its focus on wildlife conservation. The manuscript preparation was a laborious process, requiring meticulous attention to detail, adherence to strict formatting guidelines, and a thorough review of the literature. I meticulously revised and edited my manuscript many times, ensuring clarity, precision, and rigor in my writing. My advisor provided extensive guidance and feedback throughout this process, helping me refine the manuscript and strengthen my arguments.

The peer-review process itself was a unique experience. My manuscript underwent rigorous scrutiny by experts in the field, who carefully evaluated my method, data analysis, and interpretation of results. The feedback received from the reviewers was both constructive and challenging. They raised insightful questions, suggested improvements, and highlighted areas that required further clarification. I revised my manuscript multiple times in response to their comments, incorporating their suggestions and addressing their concerns. The reviewers identified a crucial oversight in my method, which meant I needed to re-analyze my data, revise my interpretations, and amend parts of my initial conclusions. This required a substantial amount of additional work, and I had to rerun the analyzes and rewrite the associated chapters of my manuscript. This was challenging and frustrating, but I gained a lot of insight into the process of rigorous scientific investigation and publication.

The absolute acceptance of my manuscript was a momentous occasion. The publication marked a significant achievement in my academic career — a testament to years of hard work, dedication, and perseverance. My research was now readily available to the global scientific community, contributing to a growing body of knowledge on the effects of climate change on Arctic wildlife. The publication also elevated my profile as a researcher, opening new opportunities for collaboration and further research.

Beyond the immediate impact of publication, my research influenced policy and conservation efforts. Various reports and policy documents cited my findings, which highlighted the vulnerability of Arctic foxes to climate change, raising awareness of the challenges facing these animals and promoting the need for conservation action. My work helped to shape conservation strategies in the Arctic region, directing resources and efforts towards protecting this vulnerable species and other Arctic wildlife. The feeling of satisfaction and purpose that arose from these actions were immense.

My contributions extended beyond individual publications. I engaged in science communication, presenting my research findings to various audiences, including students, policymakers, and the public. Through these avenues, I strived to translate complex scientific concepts into accessible language, making my research more engaging and impactful. Public lectures, interviews, and articles in popular science magazines helped to raise public awareness of climate change and its effects on wildlife. The ability to communicate scientific knowledge to a broader audience became an integral part of my work, extending the reach of my research and fostering a greater appreciation for the natural world.

My research also paved the way for future projects. Insights from my doctoral study inspired further investigations into the ecological effects of climate change in the Arctic, leading to invitations to collaborate with researchers worldwide. This expanded network laid the foundation for grant applications and broadened the scope of my work.

My journey towards completing my doctorate wasn't simply a matter of achieving an academic goal; it was a transformative experience that profoundly shaped my personal and professional life. The conference presentations and publications were not just milestones in my educational journey, but reflections of my passion for wildlife conservation, my dedication to rigorous research, and my commitment to sharing my knowledge with the world. The rewards went far beyond academic achievement. They included the personal fulfillment of contributing to a greater understanding of the natural world and a deep sense of purpose in striving to protect it.

Chapter 19: Mentorship and Collaboration Working with Experts

The publication of my research marked a significant turning point, but the journey didn't end there. It transitioned into a new phase, one characterized by collaboration and mentorship on a scale I hadn't previously experienced. The success of my independent research had opened doors, leading to invitations to contribute to larger, more complex projects. This was where the true power of collaboration became apparent.

My first significant collaborative effort involved a multi-institutional project investigating the impact of climate change on migratory bird populations across the Arctic. This ambitious undertaking involved researchers from universities in Canada, Norway, and the United States, each bringing their unique expertise and datasets to the table. The sheer scope of the project and the collective knowledge assembled in one collaborative space humbled me. My role involved analyzing satellite tracking data, using my specialized skills in data analysis and interpretation to contribute to the broader narrative.

Working within a larger team presented a unique set of challenges. Different research styles, varying levels of experience, and diverse communication preferences required careful navigation. To ensure others understood my contributions and their seamless integration with their work, I quickly learned the importance of clear communication. Regular team meetings, both in-person and remote, proved vital in maintaining cohesion and clarity of purpose. Each meeting provided an opportunity to discuss progress, address challenges, and refine our collective approach. The process of consensus-building, which involved agreeing on methodologies and interpretations, was challenging but ultimately enriching.

The project lead, Dr. Anya Petrova, a renowned ornithologist, provided invaluable mentorship. Her experience in managing large-scale research projects, combined with her expertise in ornithology and guidance in navigating the complexities of collaborative research, was invaluable. Dr. Petrova's mentorship style was one of empowerment, fostering independence while providing timely and practical support. She was adept at delegating tasks appropriately, matching individual strengths to project needs. She encouraged open communication and built an environment of mutual respect and collaboration, valuing every team member's contribution.

Beyond the scientific aspects, the collaborative project provided an invaluable opportunity for professional development. I learned to work effectively within a diverse team, negotiating conflicting priorities and perspectives. I honed my skills in scientific communication, adapting my language and approach to suit different audiences. The experience highlighted the importance of diplomacy, patience, and clear articulation in achieving a common goal. I learned to value the diverse perspectives offered by my colleagues, recognizing the power of integrating different methodologies and approaches to solve complex problems.

One particular challenge involved integrating datasets from different sources, each with its unique format and limitations. This caused the development of innovative methods for data harmonization and quality control, a process that stretched my analytical skills beyond their previous limits. The collaborative process fostered an environment of collective problem-solving, with team members readily sharing their knowledge and expertise to overcome hurdles.

As the project progressed, I took on increasingly significant responsibilities. My responsibilities included leading a subproject focused on the impact of sea ice decline on Arctic bird nesting success. This involved not only data analysis but also the coordination of fieldwork, overseeing the efforts of a smaller team of graduate students. Managing this subproject was an exceptional learning experience. It required balancing individual tasks with the overarching goals of the larger project, ensuring the seamless integration of my team's findings. This experience significantly enhanced my leadership and organizational skills, transforming me from a relatively junior researcher into a more independent and capable leader.

Published in leading scientific journals: A series of papers from the project significantly advanced our understanding of how climate change affects Arctic bird populations, representing the project's culmination. The collaborative effort yielded a body of work that was significantly more substantial and impactful than any individual researcher could have achieved alone. The experience underscored the value of collaboration in accelerating scientific progress, producing high-quality research, and increasing the reach and influence of scientific findings.

Beyond my collaborative project with Dr. Petrova, I participated in workshops and conferences, engaging with researchers from diverse backgrounds. These interactions fostered community, sparked new ideas, shaped research directions, and led to ongoing collaborative partnerships.

One helpful collaboration emerged from a conference presentation of my earlier research on Arctic foxes. A professor from a leading university in Alaska, Dr. Benicio Ramirez, approached me afterward, asking about my findings and proposing a collaborative project investigating the impact of climate change on predator-prey interactions in the Arctic. This collaboration took a slightly unique form, involving regular video conferences, data sharing, and joint manuscript preparation. The dynamic was more fluid and less hierarchical than on the previous larger project, fostering an intense intellectual exchange and mutual respect. This project again highlighted the importance of clear and articulate communication. We relied on regular virtual meetings to stay on track, address emerging challenges, and ensure alignment of goals.

The collaborative nature of my doctoral journey profoundly shaped my understanding of scientific research. It taught me the importance of teamwork, the value of diverse perspectives, and the power of collective effort in achieving meaningful scientific breakthroughs. Leading researchers mentored me, and the collaborative environment fostered my academic, personal, and professional growth. Collaboration has provided invaluable skills and experiences, preparing me for the challenges and rewards of a future scientific research career focused on collaborative efforts to solve complex problems and advance our understanding of the natural world. The network of colleagues and collaborators I developed during my doctoral studies provided a foundation of support and opportunities that extends far beyond the completion of my degree. This interconnectedness, this collaborative spirit, is what defines the landscape of modern scientific endeavor and will undoubtedly continue to shape my future contributions.

Chapter 20: Navigating Funding and Resources: The Financial Landscape

The exhilaration of collaborative research eventually gave way to the sobering reality of funding. My journey towards a doctorate wasn't solely about scientific breakthroughs; it was a constant negotiation with the financial landscape of academic research. The romantic notion of dedicated scholars pursuing knowledge without concern for monetary constraints quickly faded as I confronted the stark realities of tuition fees, living expenses, and the ever-present need for research funding.

My initial funding came from a combination of sources. A generous scholarship from the university, awarded based on academic merit and research potential, covered a significant portion of my tuition. This was a tremendous relief, but it still left a considerable gap between my income and expenses. The scholarship, while substantial, didn't account for the cost of living in a major city, nor did it cover the essential equipment and supplies needed for my research.

To bridge the financial gap, I took on several part-time jobs, balancing lectures, lab work, research, and employment. Teaching assistant roles in undergraduate veterinary courses became my main source of income, offering invaluable experience that honed my teaching skills and reinforced my understanding of core concepts. These responsibilities also strengthened my time management and ability to juggle multiple priorities.

Besides teaching, I sought contract work. Freelancing as a veterinary medical writer allowed me to leverage my knowledge and writing skills for supplemental income. I accepted projects by reviewing scientific papers, editing medical textbooks, and crafting educational materials for veterinary schools and pharmaceutical companies. This work provided flexibility, allowing me to adjust my schedule according to my academic commitments, but also required exceptional discipline to meet deadlines while navigating the academic rigors of doctoral studies. It tested my ability to focus and maintain high-quality work under pressure—a valuable lesson in the demands of professional life.

My financial challenges weren't unique; many of my fellow doctoral candidates faced similar pressures. We shared stories, exchanged strategies, and offered each other support. We discovered the power of collective resourcefulness, learning from each other's successes and mistakes. It became a constant conversation, an ever-present undercurrent to our shared academic journey. The unspoken bond formed through these shared struggles became a source of unexpected resilience and camaraderie.

One particular incident stands out as a stark reminder of the financial precariousness of academic life. A crucial piece of research equipment — a specialized microscope vital for my ongoing project — unexpectedly malfunctioned. The repair costs were exorbitant, far beyond my immediate means. My initial panic quickly gave way to a determined approach to problem-solving. I explored every available avenue, including contacting the manufacturer, seeking grants specifically for equipment repair, and reaching out to alumni networks for potential funding opportunities. The bureaucratic maze proved frustrating and time-consuming, but I persevered. Ultimately, a combination of smaller grants and a generous contribution from a former professor enabled me to secure the funds necessary to repair the essential microscope, a testament to the power of persistence and the importance of cultivating strong professional relationships.

This showed me the benefit of planning finances ahead of time. To improve my finances, I meticulously tracked expenses, budgeted income, and sought financial counsel. I gained the ability to expect funding possibilities, adapting my research proposals to fit grants and scholarships. I moved from just dealing with financial issues as they came up to proactively planning and making the most of my resources. My strategy involved continuous opportunity seeking, diligent funding call monitoring, and persistent network expansion.

Beyond formal funding sources, I embraced the power of resourcefulness. I discovered the hidden wealth in secondhand equipment, finding affordable alternatives to brand-new laboratory supplies. The online marketplace unexpectedly became a treasure trove of inexpensive, functional equipment, significantly reducing my costs without compromising the quality of my research. This resourcefulness wasn't merely about saving money; it honed my problem-solving skills and encouraged creative approaches to resource allocation.

The financial constraints also led me to refine my research method. I prioritized cost-effectiveness in experimental design, exploring alternative approaches that minimized expenditure while maximizing scientific rigor. This compelled me to become a more efficient and creative researcher, continually seeking ways to optimize resources and amplify the impact of my work.

Managing finances alongside the rigorous demands of doctoral study was a constant juggling act. There were moments of intense stress and frustration, times when the financial burden felt overwhelming. But these challenges also forged resilience and resourcefulness, shaping my character and reinforcing my dedication to my chosen path. It was a grueling but ultimately valuable lesson in self-sufficiency, strategic planning, and the importance of building a robust support network.

I became skilled at navigating the financial landscape of academic research. I learned to advocate for my needs, communicate effectively with funding bodies, and build a network of contacts for guidance and support. Securing funding, however, was only part of the experience.

Financial struggles were more than just personal. The lack of adequate funding for doctoral studies creates an inequitable system, disproportionately affecting students from lower socioeconomic backgrounds. This disparity underscores the need for increased investment in doctoral education, ensuring that talented individuals, regardless of their financial circumstances, can pursue advanced research.

My financial journey during my doctorate was not without its moments of despair. Sometimes I questioned my ability to balance the financial burden with the academic rigors. The pressure to secure funding, to maintain a viable income, and to excel in my studies felt immense. However, these challenges ultimately instilled a level of resilience and resourcefulness I hadn't expected. It taught me the importance of carefully crafting grant applications, building strong relationships with mentors and funding bodies, and constantly seeking new avenues of support. The experience transformed me from a student primarily focused on academic achievement into a well-rounded professional capable of navigating complex financial landscapes. It underscored the importance of financial planning, budgeting, and the power of effective networking — skills that would prove invaluable in my future career.

Chapter 21: Research Setbacks and Overcoming Obstacles

The unexpected death of my primary research subject, a magnificent Alaskan Malamute named Kodiak, dealt a blow that resonated far beyond the immediate loss. Kodiak was more than just a research animal; he was a crucial component of my doctoral dissertation, the centerpiece of my investigation into a novel canine autoimmune disease. His unique genetic profile, coupled with the meticulous data collected over two years, made him irreplaceable. His passing left a gaping hole in my research, threatening to derail years of painstaking work.

The initial shock was paralyzing. Grief, mingled with the crushing weight of academic failure, threatened to consume me. The meticulously designed protocols, the countless hours spent observing and recording his health, all seemed to evaporate in the face of this unforeseen tragedy. My meticulously organized data felt incomplete, a fragmented narrative lacking its pivotal conclusion. The sheer scale of the setback was daunting, challenging the very foundation of my research. Sleep became elusive, replaced by a cycle of restless nights and emotionally draining days. Doubt crept in, whispering insidious questions about my ability to recover from such a catastrophic loss and complete my doctoral studies.

The support of my mentors proved invaluable during this period.

Dr. Anya Sharma, my principal advisor, offered unwavering empathy and practical guidance. She acknowledged the enormity of the loss while gently steering me towards a path forward. Her calm reassurance and strategic thinking provided a lifeline during those dark days. She suggested revisiting my initial research proposal, exploring alternative methodologies that wouldn't cause a direct replacement for Kodiak. This was critical; it shifted my focus from the irreplaceable loss to the potential of adaptation and innovation.

Professor David Chen, a renowned immunologist, offered his expertise. He suggested I delve deeper into the existing data, exploring alternative avenues for analysis and interpretation. He reframed my research question for me, focusing on the insights already gathered instead of the lost data. His insightful guidance helped transform a potential failure into an opportunity to show innovative problem-solving skills. He pointed out that although the research was incomplete; it offered a rich dataset, providing a valuable foundation for new lines of inquiry.

This reframing of my research was a significant turning point. Rather than dwelling on Kodiak's absence, I focused on the substantial data already collected. I meticulously re-analyzed it, explored new statistical models, and collaborated with a bioinformatician to uncover previously unseen correlations. The process demanded intense focus, analytical precision, and pushed me beyond my comfort zone.

I started collaborating with a team of biostatisticians, learning new software and techniques to extract meaningful insights from the data. This collaborative environment was a crucial support system. Working with these experts provided not only technical expertise but also a sense of intellectual camaraderie. The shared intellectual challenge invigorated me, helping to replace despair with a focused determination. The process was challenging, demanding long hours and a steep learning curve. However, the collective brainstorming sessions and collaborative problem-solving were remarkably revitalizing, offering a powerful antidote to the initial feeling of isolation and defeat.

The initial findings from this reanalysis were surprisingly significant. By focusing on the existing data, I uncovered patterns and correlations that had previously gone unnoticed. These discoveries were unexpected and opened up exciting new avenues of research. The revised dissertation became a more nuanced and comprehensive study, demonstrating not only the resilience of the research process but also the ability to adapt and overcome setbacks through innovative thinking.

My revised method involved a comparative analysis of Kodiak's data with that of a control group of healthy Alaskan Malamutes. This allowed me to highlight the specific immunological markers associated with the disease, providing valuable insights for future research and potential therapeutic interventions. I also explored the genetic predispositions, identifying specific gene variations that could contribute to the disease's development. This expanded the scope of my dissertation, moving beyond the initial focus on a single animal to a broader understanding of the disease's genetic and immunological aspects.

The entire process of recovering from Kodiak's death and reshaping my research was a testament to the power of resilience and adaptability. It was a painful, challenging, and emotionally exhausting experience. However, it forced me to develop new skills, refine my research method, and show an innovative approach to scientific inquiry. The resulting dissertation was not only a successful completion of my doctoral studies but also a personal triumph, highlighting the ability to overcome adversity through perseverance, collaboration, and a willingness to adapt and learn from unexpected setbacks.

The process transformed not just my research but also my perspective on scientific inquiry. It taught me the importance of flexibility, adaptability, and collaboration in the face of unexpected challenges. The experience broadened my skills, pushing me beyond my comfort zone and forcing me to master new analytical techniques.

The journey proved that even from profound loss, something meaningful and impactful can emerge. It wasn't just the academic achievement; it was the personal growth, the resilience honed, and the unwavering commitment to my goals that made this experience a defining moment in my journey. The experience instilled a profound appreciation for the inherent uncertainties of scientific research and the importance of adapting and innovating in the face of unforeseen obstacles. The lessons learned became an integral part of my professional development, shaping my approach to scientific inquiry and my ability to navigate unexpected challenges with resilience and determination. It was a painful but ultimately transformative experience, underscoring the profound connection between personal resilience and scientific success.

Chapter 22: Maintaining Work-Life Balance, Personal Wellbeing

Astark realization followed the exhilarating rush of completing my revised dissertation: the relentless pursuit of my doctorate had taken a significant toll on me. The demanding schedule, pressure to produce important research, and constant mental strain left me exhausted. My body, once a finely tuned machine driven by ambition, was now protesting with exhaustion. Sleep deprivation became the norm, replaced only by short, restless periods of unconsciousness punctuated by anxiety dreams of failed experiments and missed deadlines. My immune system, previously robust, seemed to crumble under the strain, leaving me susceptible to frequent colds and debilitating headaches.

The vibrant energy that once propelled me forward felt dimmed, replaced by a pervasive sense of weariness.

It was during this period that I began to truly appreciate the importance of work-life balance, a concept I had previously dismissed as idealistic and unattainable. The intensity of my doctoral studies had consumed my life, leaving little room for anything else. My social life had dwindled, replaced by late-night study sessions and weekends spent poring over research papers. I had pushed my hobbies, once a source of joy and relaxation, to the forgotten corners of my mind. I realized that the very pursuit of my dream was threatening to undermine my health and well-being, creating a vicious cycle where the pressure to succeed further fueled exhaustion and burnout.

The turning point came during a grueling week, marked by endless data analysis and the relentless pressure of looming deadlines. One evening, while reviewing my research notes for the tenth time, I felt an overwhelming wave of despair wash over me.

The weight of my responsibilities felt suffocating, and I found myself unable to cope with the emotional exhaustion. It was in that moment, amidst the chaotic jumble of my research papers and overflowing coffee cups, that I recognized the urgent need for change.

My first step was to prioritize sleep. This seemed deceptively simple, yet it required a significant shift in my mindset. I learned to create a relaxing pre-sleep routine, which involved taking a warm bath, practicing calming meditation, and reading a good book–anything that allowed my mind to unwind before I fell asleep. This, combined with establishing a consistent sleep schedule, slowly but surely improved my sleep quality. The improvements were remarkable; more restful sleep led to increased energy levels, improved focus, and an enhanced ability to cope with the demands of my studies. Even minor changes made a difference–ensuring my room was dark and quiet, avoiding caffeine before bed, and maintaining consistent waking and sleeping times.

Next, I added regular exercise to my routine. At first, it seemed counterintuitive given my packed schedule, but I soon found that even short bursts of activity noticeably boosted my mood and energy. A brisk walk during my lunch break, a quick yoga session in the evenings, or a cycling trip on the weekends–any form of physical exertion provided a much-needed mental break from the intense focus required for my research. Exercise became my reset button, providing a much-needed release of endorphins and a sense of accomplishment separate from my academic endeavors. The physical activity improved not just my physical stamina, but also my mental clarity.

Equally important was the conscious effort to reconnect with my social life. I started scheduling regular coffee dates with friends, taking part in group activities, and making time for meaningful conversations. These interactions provided a crucial emotional support system, reminding me that my life extended beyond the confines of my research. I also actively cultivated connections with other graduate students, forming study groups and sharing experiences. This sense of community helped me to feel less isolated, providing mutual support and reassurance during times of stress. It provided a valuable outlet for expressing my anxieties and frustrations, normalizing the struggles we shared.

Mindfulness and meditation became invaluable tools for stress management. Pausing to center myself, even for a few minutes daily, helped reduce anxiety and overwhelm. I practiced meditation techniques, including deep breathing and guided exercises, to quiet racing thoughts and cultivate calm. These practices became essential for managing stress, improving concentration, handling pressure, and promoting overall well-being.

Dietary changes also played a significant role in enhancing my physical and mental health. I incorporated more fruits, vegetables, and whole grains into my diet while reducing my consumption of processed foods, caffeine, and alcohol. Regular meals provided sustained energy levels, helping to avoid the dreaded energy slumps that hindered my concentration. This wasn't a radical diet overhaul but a gradual shift towards healthier choices, mindful eating, and reducing reliance on quick, energy-sapping processed foods. It was a gentle transition that proved remarkably effective in improving my overall well-being.

Finally, I learned to set boundaries. This meant learning to say "no" to additional commitments when I felt overwhelmed, prioritizing tasks effectively, and scheduling dedicated downtime. It involved protecting my time, refusing to extend my work hours endlessly, and setting clear limits on my workload. This involved creating a schedule that incorporated my academic work while also allowing for ample rest, relaxation, and personal time. It required a significant shift in my thinking, moving from the mentality of relentless work to one of mindful productivity and prioritization. Learning to say "no" was difficult initially, but it proved invaluable in preventing burnout and maintaining a sustainable work-life balance.

Maintaining a work-life balance during my doctoral studies wasn't a passive process but an ongoing, conscious effort. It was a continuous learning experience, involving adjustments, experimentation, and constant self-reflection. There were days when I slipped back into old habits, forgetting the importance of self-care. However, the lessons I learned and the strategies I implemented instilled a new sense of self-awareness and prioritization that extended far beyond my academic pursuits.

The journey towards a more balanced lifestyle wasn't just about improving my physical and mental health; it was about discovering a deeper understanding of my limitations and resilience. It allowed me to value not just the attainment of my academic goals but the preservation of my overall well-being. Undeniably challenging, pursuiting a doctorate led to a fulfilling life beyond the laboratory and library. My journey showed me that genuine success encompasses navigating life's challenges holistically, learning from them, and growing professionally and personally, not just achieving goals. Life's lessons shaped my approach to life, emphasizing mindful ambition and well-being. My doctoral experience continues to shape my approach to life's challenges and triumphs. The experience transformed me, teaching me the essential balance between ambition and self-care and paving the way for a more fulfilling and sustainable future.

Chapter 23: Navigating Personal Relationships: Support and Understanding

The intense focus required for my doctoral research had cast a long shadow over my relationships. My once vibrant social life had withered, replaced by a relentless cycle of study, research, and sleep deprivation. The friends who had been constant companions throughout my undergraduate years felt increasingly distant, with their calls and texts going unanswered and their invitations politely declined. The guilt gnawed at me, a constant companion to my exhaustion. I knew they understood the pressures of my studies, but their understanding couldn't erase the feeling of letting them down, of sacrificing our connections on the altar of ambition. The distance created a quiet ache, a hollow space in my life that only intensified the loneliness of my all-consuming pursuit.

My supportive family noticed the change. My parents, always my cheerleaders, expressed a mix of pride and concern. They understood the dedication my path required but also saw the toll it was taking. My apologies and promises to make time later, frequently unfulfilled, met their attempts to engage me at family dinners. The weight of my guilt deepened, layering itself upon the already heavy burden of my studies. I felt like I was failing them, too, pushing aside the very people who had always been my constant source of strength and support.

One evening, during a rare phone call with my younger sister, the dam finally broke. Her gentle concern and quiet observation of my absence triggered an outburst of suppressed emotions. The tears flowed freely, washing away months of pent-up exhaustion and guilt. I poured out my anxieties, my fear of failure, my overwhelming sense of loneliness and isolation. Her calm voice and understanding words were a lifeline, a soothing balm to my frayed nerves. That conversation was a turning point. It forced me to acknowledge the importance of rebalancing my life and to recognize that genuine success involves more than just academic achievement. From that moment on, I attempted to reconnect with my family and friends. It wasn't easy. The demands of my research remained relentless, but I learned to prioritize those connections, to treat them as essential elements of my overall well-being, not as mere distractions from my work. I started scheduling regular calls with my sister, carving out time for video chats with my parents, and actively planning outings with my friends, no matter how small. These interactions, however brief, were a vital source of emotional nourishment, providing me with the encouragement and understanding I desperately needed.

Rebuilding those relationships involved more than just scheduling time; it also required genuine effort and attention. It required active listening, authentic engagement, and a conscious effort to be present in those moments, rather than allowing my mind to wander back to my research. I had to learn to disconnect, to truly put aside the pressures of my academic pursuits and focus on the people who mattered most. It was a conscious decision to let go of the guilt and embrace the joy of connection, to acknowledge that these relationships were not obstacles to my goals, but essential fuel on my journey.

One particular memory stands out from this time of healing and reconnection. My parents took a long weekend to visit me. They brought my favorite home-cooked meals and spent hours just talking, listening, and sharing stories. There was no pressure to discuss my research, only an unspoken understanding of the need for shared comfort and support. During one of our walks, my father shared a story from his own professional life, describing a period when he had faced similar pressures and overwhelming work demands. He spoke about seeking help when needed, about the significance of maintaining perspective and prioritization, and about the vital role of family and friends in navigating challenging times. It was during that walk that I truly understood the depth of his own experiences and the wisdom he shared. His words reassured me that my struggles were not unique, and his support reaffirmed the unwavering love and understanding that anchored me. His perspective, shared within the context of our family bond, made the challenges I faced seem less daunting, less isolating. The simple act of sharing a quiet moment amidst the stunning landscape created a space for reflection, reassurance, and emotional replenishment.

My relationship with my friends also underwent a transformation. We didn't magically erase the lost time, but we consciously rebuilt our connections. It was a gentle process, marked by quiet coffee dates, shared meals, and unhurried conversations. It was about rekindling the laughter, the shared experiences, and the unbreakable bond that had sustained us throughout our undergraduate years. These rekindled connections provided a sense of normalcy, a reassuring counterpoint to the often intense and isolating world of doctoral research. They served as a reminder that my life extended far beyond the confines of the laboratory, reminding me of the profound importance of meaningful friendships and the enduring power of human connection.

I learned that nurturing my relationships wasn't just about prioritizing time; it was about prioritizing empathy and understanding as well. It involved consciously cultivating genuine connections, practicing active listening, and truly valuing the contributions my loved ones made to my life. The support they offered wasn't just emotional; it was practical. They assisted me with household tasks, provided meals, and offered words of encouragement when I was struggling. This support enabled me to concentrate my energy on my research and approach my studies with renewed vigor and focus. It became apparent that the support system I had cultivated wasn't merely a source of emotional comfort, but a vital component of my success. It was a crucial element in maintaining my mental and emotional well-being, allowing me to persevere through the intense demands of my doctoral program.

At times, the pull of research tempted me to isolate myself, but the support and connection I experienced reinforced the importance of balance. It reaffirmed that our bonds with family and friends are the foundation of our strength and well-being.

The challenges I faced in navigating the delicate balance between my ambitions and my relationships were profound. Yet, the lessons I learned were even more valuable. I realized that genuine success encompasses more than just the attainment of academic goals; it requires a harmonious balance between our personal and professional lives, an understanding that our relationships are not distractions from our ambitions but fundamental pillars of our overall well-being. The support I received from my family and friends was not merely a source of comfort; it was the very cornerstone of my ability to persevere, to overcome obstacles, and ultimately to achieve my dream.

The lessons of balancing ambition and connection would remain with me, guiding my future endeavors and reaffirming the invaluable role of human relationships in achieving a fulfilling and purposeful life. The irreplaceable roles these relationships played in shaping my character and enabling my success powerfully show that actual achievement involves more than individual effort.

Chapter 24: Moments of Self-Doubt: Building Resilience

Acreeping sense of unease quickly tempered the exhilaration of achieving my doctorate. The mountain I had climbed, the years of relentless dedication, now felt strangely... hollow. The celebratory champagne tasted flat, the congratulations echoing with a dissonance I couldn't quite place. It wasn't a lack of accomplishment; it was a profound, unsettling self-doubt that had settled in, a shadow lurking in the corners of my newfound success.

This wasn't the triumphant feeling I'd expected. Instead, a wave of insecurity washed over me. Had I truly earned this? Was I merely lucky, a beneficiary of circumstance rather than a testament to hard work and talent? The whispers of impostor syndrome, those insidious doubts that gnaw at the edges of confidence, gained volume. I questioned my capabilities, my judgment, even my very right to hold the title I'd worked so hard to get.

The self-doubt manifested in various ways. Simple tasks, once approached with confidence, now felt fraught with uncertainty. I found myself second-guessing my clinical diagnoses, questioning my treatment plans, and even hesitating before performing routine procedures. The fear of making a mistake, of being exposed as a fraud, became a paralyzing force, threatening to undermine my professional competence.

The late nights spent poring over textbooks, once fueled by ambition, now felt like a desperate attempt to prove my inadequacy. Each unanswered question, each challenge overcome, only reinforced the nagging voice in my head, whispering that I wasn't good enough, that I would fail. This relentless self-criticism fueled a cycle of anxiety and exhaustion, draining my energy and stifling my enthusiasm.

My supportive network — the very foundation of my previous resilience — felt distant, a comfort I couldn't fully embrace. The pervasive self-doubt that threatened to overshadow every success muted the joy of sharing my accomplishments. My family's pride felt like a heavy burden. Their unwavering belief in me was a stark contrast to the chasm of insecurity that gnawed at my soul. My potential failure and the revelation of my perceived inadequacy filled me with the fear of disappointing them.

During a particularly stressful week, while struggling with a complex case, the weight of self-doubt nearly crushed me. Sleepless nights fueled by anxiety and caffeine left me exhausted, and I withdrew from the very support system that had been my lifeline.

The turning point came unexpectedly during a quiet evening spent reflecting on my journey. Instead of focusing on my perceived shortcomings, I deliberately shifted my perspective. I listed all the challenges I had overcome, all the obstacles I had navigated, and all the successes I had achieved. It was a lengthy list, a testament to my resilience, dedication, and unwavering commitment to my chosen path.

I realized that self-doubt was not a sign of incompetence, but a natural byproduct of pushing boundaries and striving for excellence. It was a reminder that the path to success is rarely straightforward, that challenges and setbacks are inevitable parts of growth. The critical voice in my head, rather than a judgment of my worth, was merely a reflection of the high standards I set for myself.

This realization shifted my perspective profoundly. My approach to self-doubt was not to vanquish it as an enemy, but to understand and manage it as a challenge. I started using strategies to combat negative self-talk, replacing critical thoughts with affirmations highlighting strengths and accomplishments. I practiced mindfulness techniques, focusing on the present moment and releasing anxieties about the future.

I also sought therapy professionally, addressing the causes of my insecurity and providing coping mechanisms. The therapist's insights helped me to recognize patterns of negative thinking and to develop healthier coping strategies. It was a process of self-discovery, involving an understanding of the roots of my self-doubt and the cultivation of self-compassion.

Slowly but surely, my confidence returned. The fear of failure lessened, giving way to a more measured approach to challenges. Embracing imperfections became my practice, understanding mistakes' role in growth. I began celebrating small victories, acknowledging my progress and celebrating the milestones along the way.

My support network was also re-engaged as I sought encouragement from family. Their unwavering belief in me helped me to recognize my worth, their perspective grounding me in reality and reminding me of my successes. The simple act of sharing my struggles and vulnerabilities strengthened my relationships and reaffirmed the importance of human connection in navigating life's challenges.

The journey towards overcoming self-doubt was not linear. Sometimes the insidious whispers returned, when feelings of insecurity threatened to overwhelm me. But with each recurrence, I approached it with greater understanding and resilience. I had developed tools to combat the negative self-talk, to manage my anxiety, and to maintain perspective.

The triumph over self-doubt wasn't about eradicating negative thoughts, but about transforming my relationship with them. It was about acknowledging their presence without letting them dictate my actions or define my self-worth. It was about embracing imperfection, learning from mistakes, and celebrating successes, no matter how small.

My journey to becoming a veterinarian was not solely about mastering the science of animal care; it was about mastering myself, about cultivating resilience, about learning to navigate the challenging emotional landscape of self-doubt. My transformed relationship with insecurity, turning a potential obstacle into a springboard for growth and achievement, came from the unwavering support of my family and friends, combined with self-reflection and professional guidance. It was an experience that reinforced this invaluable truth: genuine success is in cultivating inner strength and belief. The journey, with its challenges and breakthroughs, had shaped not only my professional life but my very essence, forging a resilience and self-belief that would serve me well throughout my career and beyond.

Chapter 25: Seeking Mentorship Guidance and Support

The initial wave of self-doubt, though terrifying, proved to be a catalyst for growth. It forced me to confront my vulnerabilities and seek the very thing I had, in my self-sufficient ambition, previously avoided: mentorship. I recognized the irony; this wasn't a sign of weakness, I realized, but a sign of maturity, a recognition that even the most driven individuals benefit from the wisdom and experience of others.

My first significant mentor was Dr. Eleanor Vance, a renowned veterinary surgeon whom I met during a summer internship. Dr. Vance wasn't just technically brilliant; she possessed an uncanny ability to connect with animals and their owners, a skill I desperately wanted to emulate. During that internship, I witnessed firsthand her compassionate approach to patient care, her unwavering dedication, and her remarkable ability to remain calm and composed even under intense pressure. She saw something in me, a spark of potential that I, clouded by self-doubt, hadn't fully recognized in myself.

One particular incident stands out. I was assisting Dr. Vance during a complex surgery on a critically injured golden retriever. Mid-procedure, I fumbled with an instrument, my hands shaking with nerves. Instead of reprimanding me, Dr. Vance calmly and patiently guided me through the steps, correcting my technique without making me feel inadequate. Her words, calm yet firm, were incredibly reassuring: "It's okay; everyone makes mistakes. The key is to learn from them and move forward. Focus on your breathing, and let's try again." Her encouragement gave me the strength to regain my composure, and we completed the surgery.

Beyond the technical skills, Dr. Vance's mentorship extended to the emotional aspects of veterinary medicine. She shared her own experiences with burnout and self-doubt, highlighting the importance of self-care and maintaining a healthy work-life balance. Her honesty, vulnerability, and willingness to share her struggles had a profoundly impactful effect. She taught me that seeking help wasn't a sign of weakness but a testament to strength and self-awareness, a crucial lesson that I would carry with me throughout my career. Her guidance helped me to cultivate a sense of resilience that extended beyond the operating room.

My next significant mentorship emerged unexpectedly from Professor Alistair Reed, my dissertation advisor. Professor Reed, a renowned expert in equine medicine, had rigorous standards and demanding expectations. Initially, I found his critiques intimidating, and my self-doubt amplified tenfold. However, as I continued working with him, I appreciated his discerning eye and unwavering commitment to excellence.

Although Professor Reed's method was difficult at first, it ended up being extremely useful. While he frankly addressed my shortcomings, his feedback was always constructive and intended to help me grow. He pushed me to think critically, challenge assumptions, and expand my horizons. He urged me to research thoroughly, hone my arguments, and present my findings with precision and clarity. I understood he didn't intend his high expectations to discourage me, but to help me reach my full potential. He helped me transform my self-doubt into a driving force for growth, transforming my apprehension into ambition.

One specific example perfectly encapsulates Professor Reed's mentorship. Methodological flaws and a lack of cohesive argument filled my initial dissertation draft. Professor Reed's critique was thorough and, frankly, devastating. I felt my confidence crumble. However, instead of dismissing my work, he spent hours patiently discussing my shortcomings and offering guidance on how to improve. He provided me with additional readings, suggested alternative methodologies, and helped me structure my arguments more effectively. He didn't offer straightforward answers; instead; he challenged me to work through the problems myself, fostering independence while providing consistent support.

This mentorship extended beyond the academic realm. Professor Reed's insight into the complexities of the veterinary profession was invaluable. In his talk, he recounted his experiences in navigating the complexities of professional politics, offering insights into effective strategies for managing challenging colleagues and resolving ethical dilemmas that often arise in such environments. Building strong relationships with colleagues, fostering an environment of collaboration, and emphasizing mutual respect were key points he underscored as crucial for success. Emphasizing the significant role of continuous learning and professional growth, he urged me to remain abreast of the latest advancements in my field, cultivating a persistent sense of curiosity and adaptability that will serve me well throughout my professional journey.

These two mentorships, though distinct in their style and approach, shared a common thread: a genuine belief in my potential and a willingness to invest their time and expertise in my development. They didn't simply advise; they cultivated my strengths, challenged my weaknesses, and helped me develop the resilience and self-belief necessary to navigate the challenges of veterinary medicine.

The importance of mentorship extended beyond these two pivotal relationships. Throughout my journey, I received guidance from many other individuals–professors, colleagues, experienced veterinarians, and even some surprisingly insightful fellow students. Each encounter, each piece of advice, and each shared experience contributed to my professional and personal growth.

The advice I received was diverse and far-reaching. Some mentors offered practical tips on specific techniques or procedures. Others shared insights into the business side of veterinary practice, providing invaluable advice on managing finances, marketing, and client relationships. Still others offered crucial emotional support, helping me navigate complex personal or professional challenges. These diverse mentorships proved incredibly helpful. Their variety helped me to see problems from multiple perspectives, enriching my experience.

Looking back, I realize that the journey to becoming a veterinarian wasn't simply a linear progression of academic achievements. It was a complex interplay of hard work, perseverance, self-doubt, and the unwavering support of mentors who invested in my growth. These mentorships weren't just about acquiring technical skills or theoretical knowledge; they were about cultivating self-belief, resilience, and the emotional intelligence necessary to thrive in a demanding profession. The challenges I overcame and the breakthroughs I achieved were all deeply intertwined with the guidance and support I received from those who believed in my potential, even when I struggled to believe in myself. The ongoing mentorship relationship continues to affect my current practice and my overall professional satisfaction. It's a testament to the enduring value of human connection and the power of shared experience. We are still far from the end of this long and difficult journey, and there are still many challenges ahead. It is the collective impact of the lessons we have learned and the persistent support we continue to receive that truly shapes and defines the trajectory of our careers.

Chapter 26: The Dissertation Process: Writing and Refinement

I have completed my Doctor of Veterinary Medicine, and my future aspiration is to further my education by pursuing a PhD that is focused on the area of animal research. Considering the extensive research already completed, as well as my prior professional experience. The next step that I must take in my academic journey is the completion of my dissertation. As the dissertation drew nearer, I acknowledged it was an enormous, monolithic task, one which, with every passing day, appeared to become more daunting and difficult. At first, the overwhelming scale of the project caused me to feel paralyzed and unable to proceed. After months of careful work to craft my research plan meticulously, it now felt flimsy and inadequate, as if it would not be enough to complete the monumental task ahead. The vastness of the unknown stretched before me like a seemingly endless ocean, threatening to swallow me whole. The fear wasn't merely academic; it was existential. This was the culmination of years of hard work, a testament to my dedication, and a defining moment in my life. The pressure was immense — a constant weight on my shoulders.

My initial research phase had been a whirlwind of late nights spent in the library, poring over countless articles and journals. I devoured books, highlighting passages, and scribbling notes in the margins. The world of equine medicine, my chosen area of focus, opened up before me, revealing layers of complexity I hadn't expected. The intricacies of equine anatomy, physiology, and pathology captivated me, leading me down rabbit holes of scientific exploration. Each discovery spurred a wave of excitement, but also fueled the anxiety of keeping pace with the ever-expanding scope of my work.

Writing the literature review took a lot of work. Synthesizing existing research, identifying knowledge gaps, and crafting a coherent narrative required hours of meticulous work. I developed an intricate system of organization, using color-coded notes, digital databases, and many spreadsheets to manage the vast volume of information. Even with this meticulous approach, the task felt overwhelming, and there were moments when I questioned my ability ever to complete it. Doubt, that insidious companion, whispered in my ear, questioning my competence and casting shadows on my progress. But I pressed on, driven by a stubborn determination to succeed.

Then came the writing. At first, the words flowed freely, fueled by the excitement of my research. The joy of discovery transformed complex concepts into an engaging narrative. But as the project progressed, I realized writing was more than transferring ideas. It was a process of refinement. I shaped and reshaped arguments, honing each paragraph until it conveyed my message with clarity and precision, often rewriting a single passage multiple times in pursuit of the perfect expression.

The editing phase was equally demanding. Professor Reed, ever vigilant, returned my drafts peppered with red ink, pointing out grammatical errors, logical inconsistencies, and gaps in my arguments. Initially, his critiques felt brutal, even disheartening. My self-doubt resurfaced, threatening to derail my progress. Yet, through it all, I pressed on. His feedback, however demanding, proved invaluable. It prompted me to refine my thoughts, sharpen my arguments, and express my ideas with precision and clarity. It was a grueling, humbling process, but I came to see his feedback as a gift, a sign of his unwavering commitment to my success.

There were moments when frustration threatened to overwhelm me. The sheer volume of work, the constant pressure to meet deadlines, and the pervasive self-doubt combined to create moments of intense emotional turmoil. There were days when I felt utterly defeated, ready to abandon the entire project. But the support of my mentors, friends, and family helped to sustain me through these difficult times. Their encouragement provided the strength and determination to continue.

The most critical lesson learned during this period wasn't just about the mechanics of writing a dissertation, but also about self-discipline and resilience. I learned to manage my time effectively, to prioritize tasks, and to overcome procrastination. I also learned the importance of self-compassion, accepting that setbacks are inevitable, and that it is okay to ask for help. The journey was fraught with challenges, moments of doubt and despair, but through it all, the unwavering support of my mentors and friends gave me a sense of strength and a renewed conviction that I could, and would, finish. Ultimately, the dissertation-writing process was a transformative experience. It was not simply an academic exercise, but a crucible that tested my resilience, determination, and self-confidence. Refining my analytical skills, this required meticulous research, painstaking writing, and rigorous editing, broadening my knowledge, and deepening my understanding of equine medicine. The result was more than just a completed dissertation; it was a testament to the power of perseverance and the transformative effect of mentorship. The feeling of accomplishment was immense, a deep-seated satisfaction that transcended the academic achievement itself.

Beyond academic success, the dissertation journey taught me the importance of meticulous planning and organization. I had to manage my time efficiently, prioritize my tasks, and create a workable schedule to ensure that I met every deadline. The entire process involved not only writing but also critical thinking, problem-solving, and the ability to synthesize and present a significant body of work coherently and concisely. The experience significantly enhanced my time management skills, research techniques, and the overall structure and presentation of my work.

Looking back, I can see that the challenges I faced while writing my dissertation weren't just academic hurdles; they were growth opportunities. They forced me to confront my limitations, develop strategies for overcoming obstacles, and cultivate a deeper understanding of my strengths and weaknesses. The journey wasn't just about writing a dissertation; it was a profound personal transformation that shaped my character, strengthened my resolve, and prepared me for the challenges ahead.

The dissertation journey provided invaluable training in the research method. I honed my skills in identifying and evaluating relevant research, critically assessing the validity and reliability of different sources, and synthesizing information from diverse sources to build a coherent argument. This process improved my research abilities and increased my proficiency in both quantitative and qualitative research methods. In my professional life, my skills have become invaluable assets.

My communication skills improved through writing the dissertation. The need to present complex scientific concepts in a clear, concise, and engaging manner pushed me to hone my written communication skills. I learned to structure my arguments effectively, use language precisely, and present my findings in a way that was both accessible and persuasive. This experience significantly improved my writing skills and advanced my ability to communicate complex ideas to both specialized and general audiences. These skills are vital for effective communication with clients, colleagues, and the wider veterinary community.

The dissertation, then, was more than just an academic exercise; it was a pivotal point in my journey, a culmination of years of dedication and a testament to the power of perseverance, mentorship, and self-belief. It provided invaluable training in research, writing, and communication skills that continues to serve me well in my professional life. Completing this monumental task was a victory not only for my academic aspirations but also for my personal growth and development. The dissertation proved to be a stepping stone to a future filled with possibility and professional fulfillment, a testament to the power of dedication and the rewards of hard work. This dissertation powerfully shows that determination, support, and belief in one's potential overcome even the most daunting challenges. It opened doors and provided opportunities I wouldn't have otherwise had access to and truly shaped my career. The lessons learned remain relevant in my continued professional development, ensuring my path continues to be guided by the skills, knowledge, and resilience I developed throughout this process.

Chapter 27: Defense Day: Presenting the Research

The weeks leading up to my dissertation defense were a blur of nervous energy and frantic preparation. Sleep became a luxury I could barely afford, replaced by countless hours spent refining my presentation, anticipating potential questions, and rehearsing my responses. My carefully organized notes, meticulously crafted over months of research, had become my lifeline — a constant companion in the late-night study sessions that defined my existence. The weight of expectation pressed down on me, a tangible force that seemed to amplify every minor imperfection in my research. I replayed scenarios in my mind, anticipating every challenge and every potential criticism.

My advisor, Professor Reed, was an unwavering source of support during this tumultuous period. His experience and guidance proved invaluable, as he helped me refine my presentation by providing valuable feedback and answering many questions. His calm demeanor and unwavering belief in my abilities helped to quell my anxieties, instilling in me a renewed sense of confidence. He didn't just guide me through the technical aspects of the defense; he also helped me navigate the emotional turmoil, providing the support I needed to face this daunting challenge. His mentorship extended beyond the academic realm; he became a confidant, offering advice and encouragement that extended far beyond the scope of my dissertation.

The day of the defense arrived like a looming storm cloud, a blend of anticipation and dread. I arrived at the conference room, my heart pounding in my chest, my hands clammy. The committee members, a panel of esteemed professors representing diverse areas of expertise, sat in their seats, their expressions inscrutable. Their silence amplified the tension, the quiet hum of expectation filling the air. Taking a deep breath, I began my presentation, a carefully crafted narrative designed to encapsulate years of research and countless hours of dedicated work. My voice, initially trembling slightly, gradually found its strength as I delved into the intricate details of my research, articulating my findings and conclusions with passion.

The presentation itself was a rollercoaster of emotions. There were moments of exhilaration as I successfully explained complex concepts, moments of panic as I stumbled over a word or struggled to recall a specific statistic. Each slide seemed to represent a milestone in my journey, a culmination of late nights, countless revisions, and the unwavering support of my mentors and friends.

The visual aids, carefully designed to enhance my presentation, played a crucial role in conveying the essence of my research, translating complex scientific information into a digestible narrative. The data, meticulously collected and analyzed, now formed the core of my argument, providing a solid foundation for my conclusions.

Once my presentation concluded, the questioning began. The committee members, engaged in a thoughtful and robust discussion, delved into the minutiae of my research, questioning my method, challenging my interpretations, and probing the limits of my understanding. Each question felt like a mini-battle, requiring quick thinking, a clear articulation of my arguments, and a thorough grasp of the literature. There were moments when I felt myself faltering, struggling to articulate my ideas with the precision and clarity that were essential to conveying them effectively. However, I persevered, drawing on the knowledge I had gained during my research, using feedback from Professor Reed and others, and relying on my understanding of equine physiology and pathology.

Their questions were not just tests of knowledge, but opportunities to demonstrate critical thinking, adaptability, and confidence in defending my work. I learned to navigate challenging inquiries that questioned my assumptions and broadened my perspective on the implications of my research. Their insights, however challenging, were invaluable, helping me to refine my arguments and enhance my understanding of my work. The process was not just about answering their questions; it was about engaging in a critical discourse, reflecting upon my research, and pushing the boundaries of my understanding.

As the defense progressed, I felt a growing sense of confidence. The initial fear and anxiety gave way to a sense of empowerment, an understanding that I was more than capable of handling this challenge. The meticulous preparation, the relentless support of my mentors, and the unwavering belief in my abilities had prepared me for this moment. My responses became more assured, my articulation more precise, and my confidence more palpable.

Finally, after what felt like an eternity, the questioning came to a close.

The committee members convened in a brief huddle, their discussion muffled but intensely focused. The anticipation was almost unbearable, the silence heavy with unspoken judgment. Then, the chair announced the decision. The committee approved my dissertation.

The wave of relief that washed over me was profound and overwhelming. Years of hard work, countless hours of research, and many moments of doubt and despair culminated in this single moment of triumph. The feeling of accomplishment was immense, a deep-seated satisfaction that transcended the academic achievement itself. This wasn't just the culmination of my educational journey; it was the culmination of a personal odyssey, a testament to the power of resilience, perseverance, and the unwavering support of mentors and friends.

Beyond the immediate joy of a successful defense, the experience provided a perspective that reshaped my view of things, including the dynamics of scholarly discourse, the art of persuasive argumentation, and the importance of rigorous self-assessment. The process refined my critical thinking skills, enhanced my communication abilities, and deepened my understanding of my field. It wasn't just about the final judgment; it was about the growth, learning, and development that occurred throughout the process.

The defense itself was a critical event, a pivotal moment in my academic journey. Still, it was also a celebration of self-discovery, a recognition of the hard work and dedication that went into achieving this milestone. It validated the journey, highlighting not just the destination but also the transformative experiences along the way. More than a formality, the defense was. Beyond my dissertation's details, the lessons expanded. The defense was a decisive step towards shaping my future and fulfilling my dreams in veterinary medicine. The memory of that day remains etched in my mind as a symbol of triumph, perseverance, and the rewards of hard work.

Chapter 28: Graduation Day: A Moment of Celebration

The crisp morning air held a hint of spring, a welcome change from the relentless winter that had shadowed the last months of my dissertation. The graduation ceremony felt less like a formal event and more like a collective exhale, a shared release of tension after years of relentless academic pressure. As I walked across the sprawling campus, the familiar buildings seemed to glow with a newfound warmth, reflecting the excitement bubbling within me.

Everywhere, I saw faces illuminated by a mixture of pride, relief, and the bittersweet anticipation of the future. Friends, classmates, and professors, all united by this shared milestone, exchanged smiles and congratulations, their conversations a vibrant tapestry of reminiscences and hopeful aspirations.

The gown, initially stiff and unfamiliar, now felt like a second skin, a symbol of the transformation I had undergone during my years of study. The weight, initially a burden, now seemed to hold the weight of my accomplishments. Each carefully folded pleat held a memory, a late-night study session, a triumphant moment of discovery, a frustrating setback overcome. As if a crown, the mortarboard perched precariously atop my head represented years of dedication and perseverance.

A carefully choreographed ballet of academic tradition, the ceremony itself was a blur of pomp and circumstance. The speeches, though eloquent and well-intentioned, faded into the background as my focus remained fixed on the anticipation of the moment I would receive my doctorate. The rhythmic cadence of the names called out echoed through the auditorium, each announcement punctuated by a ripple of applause, a wave of collective affirmation, and shared joy. I watched as my classmates, many of whom had become close friends throughout our shared journey, walked across the stage, their faces alight with pride. Each step they took felt like a symbolic victory, not just for them, but for all of us who had navigated the challenging path towards this achievement.

In the air, there was a palpable energy, a crackling feeling that was very noticeable and intense. As we looked out at the audience, the warmth and support of the families and friends, whose faces shone with pride and happiness for their loved ones, enveloped us in a powerful wave of emotion. More than a simple show of support, their presence served as a powerful statement, a visible testament to their belief in the cause. I saw my parents, their eyes welling up with emotion, and felt a surge of gratitude for their unwavering faith in my abilities and their steadfast support during times of doubt.

Then it was my turn. As they announced my name, a hush fell over the auditorium, broken only by my pounding heart. Walking across the stage felt surreal, a moment suspended between the past and the future—the years of dedication, sacrifice, and relentless hard work condensed into a single, momentous stride. As I extended my hand to receive my diploma, the weight of the document felt surprisingly light, a stark contrast to the years of responsibility and commitment that had led up to this moment. The Dean's handshake, though brief, was a powerful symbol of recognition, a formal acknowledgment of my accomplishment.

The feeling was indescribable — a potent cocktail of relief, joy, and overwhelming gratitude. Tears welled up in my eyes, blurring my vision, as I looked out at the sea of faces, my heart swelling with emotion. It wasn't merely the academic achievement that evoked such powerful feelings, but the culmination of a personal odyssey — a journey of self-discovery, resilience, and growth. The diploma was not just a piece of paper; it was a testament to the countless hours of study, dedication, and the unwavering support of my mentors, as well as the many sacrifices I had made along the way.

The post-ceremony celebrations were a joyous blur of laughter, hugs, and shared stories. Friends and family gathered, recounting their favorite memories from our time together, celebrating not just the academic accomplishments but also the bonds forged through shared experiences and mutual support. The campus, transformed into a scene of jubilant chaos, reverberated with the sounds of laughter, music, and heartfelt conversations. The air was alive with the energy of accomplishment and the anticipation of the future.

There were celebratory speeches, heartfelt toasts, and an overwhelming outpouring of love and support. A profound sense of connection and a shared understanding of our collective journey infused every interaction. The feeling of camaraderie was palpable, an unspoken acknowledgment of the common ground shared by those who had persevered through the rigorous academic demands of veterinary medicine.

Intimate gatherings filled the evenings, as small groups of friends and family celebrated each other's successes. Friends and family shared memories, told stories, and created an atmosphere of profound gratitude and joy. The focus wasn't just on individual accomplishments, but on the collective journey and the support system that had propelled each graduate towards success. The conversation transitioned seamlessly from academic achievements to personal triumphs, reflecting the holistic nature of the transformation we had all experienced.

Both liberation and responsibility: I felt these in the following days. The weight of expectations eased, yet the weight of responsibility increased as I looked towards the future and the professional opportunities that lay ahead. The realization of the responsibility that comes with being a Doctor of Veterinary Medicine tempered the freedom to pursue my dreams.

Looking back, the graduation ceremony not only recognized academic achievement but also celebrated perseverance, resilience, and unwavering self-belief, marking profound personal growth. It was a moment of profound gratitude, not only for the opportunity to pursue my passion but also for the steadfast support of my family, friends, mentors, and professors who helped guide me along the way. It was the culmination of a chapter, but more importantly, the beginning of a new and exciting one, full of possibilities and the promise of making a meaningful difference in the lives of animals and their people. The memory of that day, etched into my mind, serves as a constant reminder of the power of hard work, dedication, and the unwavering support of those who believe in you. The joy and sense of accomplishment remain vivid, a constant source of inspiration and motivation as I embark on the next stage of my professional journey. It was in every sense the culmination of a dream, a dream that continues to unfold with each passing day.

Chapter 29: Post-Graduation Reflections Looking Back

Lost in thought at the window, I forgot my half-empty mug of lukewarm tea in the small kitchen, where only the quiet hum of the refrigerator broke the silence. The vibrant tapestry of the graduation ceremony, the joyous chaos of the celebrations, felt like a distant dream, a vivid memory fading into the softer hues of everyday life. The crisp edges of accomplishment were slowly blurring, replaced by the familiar, comforting softness of routine. But the quiet contemplation wasn't melancholic; it was a quiet joy, a deep satisfaction that resonated within me.

It had been a long road, paved with late nights hunched over textbooks, fueled by countless cups of coffee and the unwavering support of my family and friends. I recalled the early days, the nervous energy of my first veterinary science competition, the thrill of victory, and the sting of defeat. Each experience, no matter how small, had served as a stepping stone, shaping my skills and my resilience. The countless hours spent tending to injured animals, the endless research papers, the grueling exams–each challenge had honed my abilities, sharpening my focus and deepening my understanding of the intricate world of veterinary medicine.

A vivid memory lingered: the dissertation, the product of years of focused research. At first glance, the task seemed insurmountable because of its sheer magnitude—an overwhelming mountain of data and analysis. There were days filled with profound frustration, days when the weight of expectation threatened to crush me. But amidst the uncertainty and doubt, I discovered an inner strength I never knew I possessed. The process wasn't merely academic; it was a crucible that forged my character and shaped my identity. It taught me the value of perseverance, the importance of seeking guidance, and the power of collaboration.

I learned to embrace setbacks, viewing them not as failures, but as opportunities for growth. Every research dead end, every unexpected obstacle, pushed me to think critically, innovate, and develop a more robust and nuanced approach to my work. The support of my advisor, Dr. Evans, was invaluable during this period. His guidance, along with his unwavering belief in my capabilities, was a lifeline during turbulent times. His gentle encouragement and insightful feedback helped me navigate the complex landscape of research and refine my arguments. He taught me not just the technical skills necessary for scientific inquiry, but also the importance of clear communication and the value of a rigorous approach to problem-solving.

The relationship I forged with my classmates was also an integral part of my journey. The shared struggles, the late-night study sessions, the mutual support—these experiences created bonds that would last a lifetime. They were not just fellow students; they were my friends, my confidantes, my sounding board during times of stress and uncertainty. Their unwavering encouragement and their unwavering belief in my abilities gave me the strength to persevere, even when the task seemed insurmountable. I remember Sarah — ever-optimistic and unfailingly supportive, always ready with a word of encouragement and a comforting laugh. And Mark, the quiet, contemplative one, whose insightful comments often provided the clarity I needed to overcome a tough challenge. These friendships, forged in the crucible of shared academic experience, were a source of constant strength and inspiration.

But academic pursuits did not solely define the path. I intertwined my journey to my doctorate with a concurrent journey of self-discovery. I learned to recognize and challenge my limitations, to step outside my comfort zone, and to embrace vulnerability. The process wasn't just about acquiring knowledge; it was about developing the skills and confidence to navigate life's complexities, both professionally and personally. I learned the importance of self-care, maintaining a healthy work-life balance, and recognizing the value of seeking support during times of stress.

The self-doubt that once plagued me gradually gave way to quiet confidence. I learned to trust my instincts, stand by my convictions, and value my perspective. This self-assurance wasn't just the result of academic success, but of years of hard work, perseverance, and reflection.

Looking back, the challenges weren't just obstacles to be overcome; they were essential components of my growth. The frustration, the setbacks, the moments of self-doubt–these experiences shaped my character, deepening my empathy, honing my problem-solving skills, and fostering a resilience that I carry with me to this day. The journey taught me the value of patience, the importance of perseverance, and the power of self-belief.

The recognition was significant, extending far beyond the typical formal academic accolades. That moment didn't signify an end, but a commencement, a platform from which to launch into uncharted territories filled with fresh challenges and exciting opportunities. As I gazed into the horizon, I saw the vast expanse of the future unfolding before me, promising endless potential and countless options, showcasing how dreams can materialize with determination, hard work, and a firm faith in one's own skills.

The world of veterinary medicine awaited its challenges and rewards, both daunting and inspiring. The experiences of the past few years have equipped me not only with the knowledge and skills necessary to excel in my chosen field, but also with the resilience and confidence to navigate life's complexities. Challenging, demanding, and overwhelming, the journey proved transformative, revealing unexpected strengths and capabilities. It was a journey that transformed a young, aspiring veterinarian into a confident, compassionate, and dedicated professional, ready to face the challenges and embrace the rewards of a life devoted to the care of animals. And that realization, perhaps more than the doctorate itself, was the greatest reward of all. Hanging on my wall, the carefully framed diploma symbolizes my achievement and reminds me of the incredible journey I'll cherish forever. The late nights, triumphs, setbacks, and friendships shaped me and who I would become, becoming interwoven into the fabric of my being. The future beckoned, clear and filled with the promise of recent adventures and a fulfilled dream.

Chapter 30: Future Aspirations Looking Ahead

Years of relentless effort culminated in the framed diploma hanging on the wall, silently observing my quiet contemplation. The weight of accomplishment had settled, replaced by the exhilarating lightness of possibility. The future, once a distant, hazy horizon, now stretched before me, vibrant and full of promise. My doctorate wasn't a finish line; it was a launchpad, propelling me towards a future brimming with opportunities to contribute to the field I loved.

My immediate plans were straightforward, yet deeply satisfying. I got a postdoctoral fellowship at the renowned veterinary research center in Davis, California. Working with leading veterinary cancer experts on important research was both exciting and an enormous responsibility. I envisioned myself immersed in the rigorous demands of the lab, the intellectual stimulation of collaborative research, and the thrill of making discoveries. The fellowship wouldn't just hone my research skills; it would also expand my professional network, connecting me with a community of like-minded individuals driven by a shared passion for advancing veterinary medicine.

Beyond the fellowship, my aspirations soared. My dream was to start an animal hospital that combined compassionate care with advanced technology. Not just any practice, but a clinic focused on holistic animal care, integrating traditional veterinary medicine with complementary therapies, creating a space where the physical and emotional well-being of each animal was paramount. I envisioned a team of dedicated professionals, united by a shared goal: to provide exceptional care and build trusting relationships with every client. The clinic would be more than a place of healing; it would be a sanctuary, a space of comfort and reassurance for both animals and their human companions.

The design of the clinic itself held a special place in my imagination. It would be a welcoming, stress-free environment, designed to minimize anxiety for animals. Soft lighting, calming colors, and spacious examination rooms would create a tranquil atmosphere. Play areas for animals awaiting treatment would add an element of fun, reducing the apprehension often associated with veterinary visits. The clinic would also incorporate innovative technology, ensuring that my patients received the most advanced and effective treatments available. I imagined state-of-the-art diagnostic equipment, minimally invasive surgical techniques, and a commitment to ongoing professional development for the entire staff.

My commitment to education wouldn't end with my doctorate. I planned to engage actively in continuing education, staying abreast of the latest advances in veterinary medicine. Attending conferences, taking part in workshops, and engaging in self-directed learning will be integral to my professional growth. I saw this not as an obligation, but as an opportunity for continued learning and self-improvement. The ever-developing nature of veterinary medicine demanded constant learning, and I embraced this challenge with enthusiasm. Staying at the forefront of my field was not merely a matter of professional pride; it was a commitment to delivering the highest quality of care to my patients.

The impact I longed to make extended beyond individual patient care. I envisioned myself playing a significant role in educating future generations of veterinarians. Mentoring students, sharing my knowledge and experience, guiding them through the complexities of the profession–these were goals I held dear. I hoped to instill in them not only the technical skills necessary for veterinary practice but also the compassion, empathy, and ethical considerations that are essential to the profession. I wanted them to approach their work with a holistic understanding of animal health, recognizing the intricate interplay between physical, emotional, and environmental factors that influence animal well-being.

My passion also extended to community engagement. I dreamed of creating and implementing outreach programs designed to promote animal welfare and educate the public about responsible pet ownership. I planned to collaborate with local shelters and rescue organizations, providing veterinary services to animals in need and advocating for their welfare. Every animal, in my belief, deserves compassionate care and a happy, healthy life regardless of breed, background, or health status. My commitment extended to educating communities about responsible pet ownership, promoting responsible breeding practices, and advocating for animal welfare legislation.

Contributing to scientific advancement remained a central pillar of my aspirations. I intended to continue my research efforts, exploring innovative treatment strategies for cancer in animals and contributing to the development of novel diagnostic tools. The pursuit of knowledge, the dedication to research, and the desire to contribute significantly to the field were driving forces in my career trajectory. I thought important research was necessary to improve veterinary medicine and the lives of animals everywhere.

But beyond the ambitious goals, I held a more profound, more personal aspiration: to inspire. I wanted to be a role model for aspiring veterinarians, demonstrating the power of passion, perseverance, and dedication. I wanted to show them that the path to achieving their dreams might be challenging, but it would be immeasurably rewarding. My journey hadn't been easy, but every obstacle I overcame, every challenge I met, had strengthened my resolve and fueled my passion. The hardships and successes had shaped me, molding me into the veterinarian I had become — someone compassionate, dedicated, and resolute in my pursuit of excellence.

My profession wouldn't solely define my future achievements; my relationships would enrich it. Family and friends, the bedrock of my support system, would continue to play a vital role in my life. I envisioned a future where my fulfillment and professional success intertwined, and where I shared the joy of my work with loved ones. The balance of work and life, once a struggle, would become a carefully orchestrated harmony, a testament to the lessons I had learned along the way.

Countless possibilities filled the beckoning future. The path ahead might be unpredictable, full of unexpected twists and turns, but I knew I was prepared. The years of study, the challenges overcome, the relationships forged–all these experiences had shaped me, equipping me with the knowledge, skills, and resilience to navigate whatever lay ahead. My journey to becoming a veterinarian hadn't merely been about earning a doctorate; it had been a transformative odyssey of self-discovery, perseverance, and the unwavering pursuit of a dream. As I looked towards the future, I carried with me not only my diploma but also the confidence, compassion, and steadfast dedication that would define my future contributions to the field of veterinary medicine. The future was a blank canvas, and I was ready to paint it with the vibrant hues of my passion and purpose.

Chapter 31: Volunteer Work Animal Welfare Initiatives

The postdoctoral fellowship in Davis proved to be everything I had hoped for, and more. The rigorous research, the collaborative spirit, and the intellectual stimulation were invigorating. Yet, amidst the demanding schedule of lab work and data analysis, a persistent feeling tugged at my conscience–a yearning to engage actively with the broader community and contribute beyond the confines of the research lab. A deep-seated passion for animal welfare had always fueled my academic pursuits, and I realized that true fulfillment lay not only in scientific advancement but also in tangible, hands-on contributions to the lives of animals in need.

This realization led me to seek volunteer opportunities that aligned with my expertise and passion. My first significant involvement was with the local animal shelter, a bustling haven for abandoned, neglected, and injured animals. Initially, I offered my services for a few hours each week, providing routine medical examinations, administering vaccinations, and treating minor injuries. However, my commitment quickly grew beyond those initial hours. I found myself drawn to the stories of each animal, their unique personalities, and their desperate need for care and compassion.

The shelter often lacked resources, and I frequently handled cases beyond routine care. I vividly recall one particularly poignant instance involving a young, severely injured stray dog named Lucky. Someone had found him abandoned, his leg broken and infected. The initial assessment revealed the need for immediate surgery, but the shelter's limited funds presented a significant obstacle. I rallied my colleagues from the research center, and together, we raised the funds through a combined effort of personal donations and an online fundraising campaign. Lucky had successful surgery, and after weeks of intensive care and rehabilitation, a loving home adopted him. His story, and countless others like it, solidified my commitment to providing veterinary care to animals in need. Beyond direct medical care, I became deeply involved in the shelter's educational outreach programs.

I developed and delivered presentations to local schools and community groups on responsible pet ownership, emphasizing the importance of spaying and neutering, vaccinations, proper nutrition, and the ethical treatment of animals. These sessions weren't just lectures; they were interactive experiences designed to engage both children and adults. I used engaging visual aids, interactive games, and real-life examples to make the information accessible and memorable. The children would often bring in their stuffed animals for "check-ups," and the adults would actively take part in discussions about responsible pet ownership and the challenges faced by animal shelters.

My volunteer work extended beyond the local shelter. I joined forces with a wildlife rehabilitation center, a place where injured and orphaned wild animals received expert care before being released back into their natural habitat. My expertise in veterinary medicine proved invaluable in treating a wide range of wildlife species, from injured birds to orphaned raccoons. This work was challenging because of the unique medical needs of wild animals and the stringent protocols required for their handling and rehabilitation. But the satisfaction of releasing a rehabilitated animal back into the wild, watching it rejoin its natural environment, was a profound and enriching experience. It emphasized the importance of conservation efforts and the crucial role of veterinary medicine in safeguarding wildlife.

Another facet of my volunteer work involved collaborating with local rescue organizations, which focused on rescuing and re-homing animals from high-kill shelters. These organizations rely heavily on volunteers to transport animals, provide temporary foster care, and assist with fundraising activities. I helped establish a mobile veterinary clinic that provided essential medical services to animals in underserved communities. The clinic offered low-cost or free vaccinations, spaying and neutering services, and general health checks to animals owned by low-income families. This initiative played a crucial role in enhancing the health and well-being of animals in these communities, while also educating pet owners about preventive care.

The experience of working with these diverse animal welfare organizations taught me invaluable lessons about teamwork, resource management, and the unwavering dedication required to make a meaningful difference in the lives of animals. I learned to adapt my veterinary skills to various settings and species, from the sterile environment of a research lab to the often chaotic yet rewarding atmosphere of an animal shelter. These experiences not only honed my clinical abilities, but also refined my organizational and leadership skills. I assumed increasingly significant leadership roles within volunteer organizations, coordinating volunteer efforts, developing outreach strategies, and managing fundraising initiatives.

One particular challenge I faced was navigating the complex emotional toll of working with animals in distress. Witnessing the suffering of neglected or injured animals could be emotionally draining, but it also fueled my determination to make a positive impact. I learned to develop coping mechanisms to manage the emotional burden, and I also emphasized the importance of self-care and mutual support among fellow volunteers.

The rewards of my volunteer work extended beyond the direct impact on animals. It profoundly enriched my understanding of the human-animal bond and the complexities of animal welfare issues. I saw firsthand the challenges faced by animal shelters and rescue organizations, the importance of community support, and the far-reaching impact of responsible pet ownership. The experiences not only reinforced my commitment to veterinary medicine but also strengthened my understanding of community engagement and the role of veterinary professionals in advocating for animal welfare.

The volunteer work became a significant part of my identity, a testament to my values and a reflection of my commitment to giving back to the community. Developing my leadership, improving my relationships, and furthering my understanding of animal welfare were all significant aspects of my personal growth. It wasn't simply about providing medical care; it was about building relationships with animals, their owners, and the broader community, fostering a deeper understanding and appreciation for the importance of compassion, empathy, and responsible stewardship of animal life.

The commitment to volunteer work extended beyond the immediate impact on individual animals. The goal was to raise awareness, advocate for improved animal welfare policies, and inspire others to join the collective effort to create a kinder, more compassionate world for animals. This work, while demanding, was also an exceptionally fulfilling aspect of my journey, providing a sense of purpose that went beyond my professional achievements. It reinforced my belief in the power of individual actions to create meaningful change, and it instilled a sense of optimism about the future of animal welfare. My commitment to this work remains unchanged to this day, and it continues to be an integral part of who I am as both a veterinarian and a person. It's a constant reminder that true fulfillment lies not only in pursuing professional goals but also in actively working to create a better world for those who depend on our care and compassion. The journey continues, and the animals and the community remain at its heart.

Chapter 32: Mentoring Aspiring Veterinarians: Sharing Experience

The satisfaction of rehabilitating Lucky and countless other animals ignited a new passion within me: mentoring. The privilege of practicing veterinary medicine felt incomplete without sharing the accumulated knowledge and hard-earned wisdom with the next generation. It wasn't just about teaching techniques; it was about fostering a love for the profession, instilling a strong ethical compass, and nurturing the empathy crucial for effective veterinary care.

My first mentee, Sarah, was a bright-eyed undergraduate student brimming with enthusiasm but also burdened by self-doubt. She'd shadowed me at the wildlife rehabilitation center, mesmerized by my interactions with the animals. Her initial anxieties were palpable—the fear of making mistakes, the pressure of handling delicate creatures, the sheer weight of responsibility. I recognized those feelings; they were echoes of my early struggles.

Our mentorship began with simple tasks—cleaning cages, preparing food, and observing my examination techniques. Gradually, I entrusted her with more responsibility, allowing her to assist with basic wound care under my strict supervision. I remember the day she successfully bandaged a fractured bird's wing. Her face, illuminated by a mixture of pride and relief, was a testament to her burgeoning confidence.

Our relationship extended beyond the clinical setting. We'd share lunch, discussing not just veterinary science but also the challenges of balancing personal life with a demanding career. I encouraged her to join the veterinary science club, take part in competitions, and network with other students and professionals. Beyond technical skills, I also imparted. We discussed ethical dilemmas, the importance of client communication, and the emotional resilience needed to cope with the heartbreak of losing a patient.

Sarah's journey wasn't without its obstacles. There were setbacks, moments of frustration, and times when her confidence wavered. But through it all, I remained a constant source of encouragement and support. I shared my failures and how I learned from them, demonstrating that mistakes are inevitable, but learning from them is essential. I showed her how to develop critical thinking skills, approach challenging cases, and work effectively as part of a team.

Years later, Sarah became a successful veterinarian, specializing in avian medicine. Her dedication and compassion mirrored the values I'd tried to instill in her. She's now a mentor herself, paying forward the kindness and guidance she received. Her success isn't just a testament to her hard work and talent; it's also a validation of the power of mentorship.

My mentorship extended beyond one-on-one relationships. I participated in formal programs within the university and local veterinary associations, providing workshops, seminars, and networking opportunities to support aspiring veterinarians. I also helped design a curriculum emphasizing practical skills, ethics, and professional development. One particularly fulfilling program focused on underserved communities, connecting students with local animal shelters and community clinics. These experiences broadened their perspectives, exposing them to the realities of veterinary medicine beyond the pristine environment of a university hospital.

One memorable experience involved mentoring a group of students during a disaster relief operation. A major hurricane had devastated a coastal region, leaving many animals injured and displaced. The students, initially overwhelmed by the chaos and the sheer number of animals requiring immediate attention, quickly adapted under my guidance. We worked tirelessly, providing emergency medical care, triage, and administering vaccines. The experience fostered a profound sense of teamwork and resilience. They learned firsthand the importance of adaptability, resourcefulness, and the ability to make critical decisions under pressure.

Mentoring wasn't always smooth sailing. Some students responded more readily to guidance than others did. Some needed more encouragement; others, more structure. Adapting my approach to meet individual needs was a constant learning process. I learned the importance of active listening, recognizing individual learning styles, and fostering a safe space for students to ask questions, voice concerns, and express doubts without fear of judgment.

Beyond formal mentorship programs, I sought opportunities to engage with aspiring veterinarians informally. I hosted regular "coffee chats" where students could drop by, share their experiences, and discuss their aspirations. These relaxed interactions fostered trust and created a supportive environment where students felt comfortable expressing themselves openly. I became a confidante, offering advice on everything from choosing a specialty to managing stress and balancing work-life demands.

My role as a mentor wasn't simply about imparting knowledge; it was about fostering a love for the profession, building confidence, and nurturing their professional and personal growth. I believe that cultivating a supportive community among aspiring veterinarians is crucial. I often encouraged my mentees to connect, providing opportunities for peer-to-peer support and collaboration.

My journey of giving back to the community isn't just about treating animals; it's about nurturing the next generation of compassionate and skilled veterinarians. It's about creating a ripple effect of kindness and expertise, ensuring that the unwavering dedication to animal welfare continues far into the future. The fulfillment I derive from mentoring is immense — a testament to the enriching power of sharing knowledge and experience, and fostering future leaders in veterinary medicine. Ultimately, this contributes to a better world for animals. The legacy isn't just in the animals I've treated but also in the lives I've touched and the future veterinarians I've inspired to make a difference. As the circle continues to expand, the impact multiplies, and that, more than anything, is the greatest reward.

Chapter 33: Community Outreach Programs Educating the Public

My commitment to animal welfare extended beyond the clinic walls and into the heart of the community. I discovered a profound need for public education regarding animal health and welfare, a realization that spurred me to take part actively and even develop several community outreach programs. Treating individual animals was only one piece of a larger puzzle; the actual impact required a shift in community understanding and behavior.

One of the most rewarding initiatives was the "Pawsitive Partners" program, a collaboration between my practice and local schools. We designed a curriculum that introduced children to responsible pet ownership. This wasn't just about feeding and walking; it delved into the complexities of animal behavior, understanding their needs, and recognizing signs of illness or distress. We held interactive workshops, complete with demonstrations of proper handling techniques, grooming basics, and even basic first aid for common pet injuries. The children responded with enthusiasm, their eyes wide with wonder as they learned about the intricacies of animal care. One particularly memorable session involved a visit from a therapy dog, showcasing the calming and therapeutic benefits of human-animal interaction. The feedback from teachers and parents was overwhelmingly positive, highlighting the program's success in fostering empathy and responsibility in young minds.

The success of "Pawsitive Partners" emboldened us to reach a wider audience. We organized community events in partnership with local animal shelters and rescue organizations. These events, held in parks and community centers, provided a platform to educate the public on various animal welfare issues. We set up informational booths, featuring posters and brochures on responsible pet ownership and up. Free micro-chipping and vaccination clinics protected pets from preventable diseases. We also organized adoption days, giving abandoned animals a chance at finding loving homes. The turnout at these events was consistently high, demonstrating a simple desire for information and a willingness to take part in animal welfare initiatives. We observed a significant increase in responsible pet ownership practices and a heightened awareness of the community's role in promoting animal safety and well-being.

Beyond these organized events, I actively sought opportunities to engage with the community through public speaking engagements. I have given presentations at schools, community centers, and local businesses, sharing my expertise and passion for animal welfare. My talks covered a broad range of topics, including recognizing the signs of animal abuse, understanding animal behavior, and promoting responsible pet ownership. The discussions that followed were often insightful and eye-opening, revealing a spectrum of public perceptions and concerns about animals. These presentations also served as valuable opportunities to address myths and misconceptions about animals, dispelling unfounded fears and promoting a more balanced and informed understanding. In one particularly moving instance, I presented to a group of farmers, highlighting the importance of responsible livestock management and addressing their concerns about animal health and welfare, while also emphasizing the importance of humane farming practices.

Recognizing the limitations of traditional methods, I embraced the power of social media to reach a wider audience. To help our patients stay informed, we've created a social media page with regular updates, tips, and educational materials. We launched online campaigns, such as "Know Your Pet's Signals," focusing on identifying subtle signs of illness or distress. We used engaging visuals and informative infographics to make complex information accessible to a broader audience. Social media also proved to be an incredibly effective tool for promoting our community events, boosting attendance and participation. This innovative approach was crucial in reaching a younger demographic and adapting our outreach to their preferred mode of communication.

I recognized the power of partnership in promoting animal welfare. We collaborated with local veterinary organizations, humane societies, and animal protection groups to coordinate our efforts. This approach ensured that our messages were consistent, impactful, and reached a larger segment of the population. We supported each other's initiatives, pooling resources and sharing expertise to maximize our collective influence. One successful example involved collaborating with a local wildlife rehabilitation center to develop an educational program for schoolchildren, which combined animal care demonstrations with interactive lessons on conservation and wildlife preservation.

The impact of our community outreach programs was measurable. We witnessed a noticeable improvement in public awareness of animal welfare issues. Animal shelters reported a decrease in the number of abandoned animals, a testament to the improvement in responsible pet ownership practices. Countless grateful community members sent us messages, expressing their appreciation for our work. It was the intangible changes, however–the growing empathy, the increased responsibility, and the enhanced community engagement–that served as the most significant measure of our success. The realization that we were not simply treating animals but were changing perceptions and transforming attitudes was incredibly fulfilling.

The experience of establishing and managing these programs was not without its challenges. Securing funding, managing volunteers, and coordinating schedules all require time, patience, and meticulous planning. There were moments of frustration and setbacks, such as low turnout at an event or unexpected logistical challenges. However, the unwavering support of our team, the dedication of our volunteers, and the enthusiastic response from the community far outweighed any difficulties.

The journey of giving back to the community has been a transformative one, teaching me valuable lessons about leadership, collaboration, and the profound power of education. This reinforced my belief that veterinary medicine extends beyond the clinic. The ripple effect of this collective effort continues to inspire me and reinforces my commitment to making a lasting difference, one paw print at a time.

Chapter 34: Fundraising Efforts Supporting Animal Shelters

My work with community outreach programs highlighted a critical need: the need for sustainable funding for animal shelters. These vital organizations, often operating on shoestring budgets, are the frontline defenders of abandoned and neglected animals. Their tireless efforts provide shelter, food, medical care, and ultimately, a second chance at a loving home for countless creatures. Witnessing their dedication firsthand ignited a fire within me—a determination to contribute beyond my volunteer hours and provide tangible support.

The first step was identifying our fundraising goals. We needed to secure sufficient funds to cover essential operational expenses, including veterinary care, food, shelter maintenance, and staff salaries. We also wanted to support specialized programs, like behavioral rehabilitation for traumatized animals or medical care for animals with chronic conditions. Setting realistic yet ambitious targets felt like a crucial first step in creating a sustainable support system. We aimed for a phased approach; initially focusing on securing enough to cover immediate necessities and then expanding our ambitions to include more long-term, programmatic goals.

Next, we needed to craft an interesting narrative. Fundraising wasn't just about numbers; it was about telling a story. We had to connect with potential donors on an emotional level, illustrating the real-life impact of their contributions. We developed a series of short videos showcasing the heartwarming transformations of animals rescued from dire circumstances—animals who, with the help of dedicated shelter staff and the community, had found loving homes. These videos, shared across various platforms, aimed to highlight the incredible work being done and the profound difference even small contributions could make.

We launched a multifaceted fundraising campaign. The cornerstone was an online crowdfunding platform that strategically used social media to maximize its reach. We created visually engaging content, including photos and videos of the animals, highlighting their unique personalities and stories. We strategically emphasized the shelter's impact on the community, not just the individual animals, underscoring the ripple effect of their work in reducing animal suffering. This broadened our appeal beyond animal lovers to include people who value community well-being.

Alongside the online campaign, we organized several fundraising events. The "Paws for a Cause" 5K run/walk has become an annual tradition, drawing participants of all ages and fitness levels. This inclusive event provided a fun, community-oriented experience, further raising awareness about the shelter's mission. We partnered with local businesses, offering sponsorship opportunities for event promotion and recognition. This collaborative approach enabled us to leverage existing networks and expand our reach, significantly boosting our fundraising efforts.

We also hosted an elegant gala, "A Night for the Animals," featuring a silent auction, gourmet food, and live entertainment. This high end event targeted potential donors who could make large contributions. We carefully curated a selection of auction items, including artwork, travel experiences, and unique animal-themed items, to create an interesting appeal for a broad audience. The gala showcased the best of our community, demonstrating a unified commitment to animal welfare.

Instead of traditional methods, we explored other options. Exploring innovative avenues, we investigated corporate sponsorships. We pitched proposals to local companies, highlighting the benefits of associating their brand with animal welfare initiatives. We emphasized the positive public relations and goodwill related to supporting a cause that resonated with a large segment of the population. This approach proved successful, with several businesses committing to long-term partnerships. Some companies even integrated our campaign into their employee engagement programs, providing matching funds for employee donations.

In addition to large fundraising events, we developed smaller, ongoing initiatives. Donation boxes in local businesses provided a simple way for regular contributions, and a monthly giving program encouraged lasting impact. These efforts created a steady funding stream, reducing reliance on sporadic events. We even implemented a "wish list" system on our website, allowing people to donate specific items needed by the shelter, addressing immediate needs such as food, bedding, or medical supplies.

Throughout the fundraising efforts, transparency and accountability were paramount. We meticulously tracked all income and expenses, publishing regular reports online to keep our donors informed of how their contributions were being used. This fostered trust and ensured the ongoing support of our generous contributors. We even organized shelter tours, allowing donors to witness firsthand the impact of their generosity on the animals and shelter staff.

One gratifying aspect of the fundraising was the community involvement. Local artists donated their work for the auctions, businesses offered discounts and services, and individuals volunteered their time and skills. This collaborative spirit significantly amplified our efforts, creating a sense of shared responsibility and accomplishment. It wasn't just about raising money; it was about building a community dedicated to animal welfare.

The success of our fundraising efforts extended far beyond the financial gains. We strengthened our connections with the community, building relationships with individuals, businesses, and organizations. This network of support became a vital asset, extending beyond financial contributions to encompass volunteer work, advocacy, and community awareness. The shared commitment to animal welfare fostered a stronger sense of community and deepened our collective impact.

The challenges, naturally, were present. We faced logistical hurdles, including coordinating volunteers for events and navigating the complexities of grant applications. There were moments of frustration when fundraising targets seemed out of reach. However, the unwavering support of our team, the passionate dedication of our volunteers, and the incredible generosity of our donors always kept us going. We learned to adapt, refine our strategies, and celebrate minor victories along the way.

Looking back, the fundraising journey was as transformative as the community outreach initiatives. It taught me the power of collaboration, the importance of effective communication, and the profound impact of collective action. It reinforced my belief in the power of storytelling and the ability to inspire others to join a cause they believe in. The financial resources secured helped to provide essential services to the animals in need, but the lasting impact extended far beyond the numbers. It fostered a stronger sense of community, a renewed commitment to animal welfare, and a shared understanding that we all have a role to play in making a difference in the lives of vulnerable animals. It is this enduring legacy that truly measures the success of our work, solidifying my conviction to continue this journey of giving back, one paw print at a time.

Chapter 35: Collaboration with other professionals: Teamwork in Animal Welfare

My experience with community outreach and fundraising underscored a critical truth: significant impact on animal welfare causes requires a collaborative approach. No single individual, however dedicated, can shoulder the immense responsibility of addressing the multifaceted needs of animals in crisis. It's the collaboration of diverse expertise and shared commitment that truly transforms the landscape of animal welfare.

This realization spurred me to seek partnerships actively with other professionals across the veterinary spectrum. I forged powerful alliances with experienced veterinarians specializing in various fields–from avian specialists who could guide us on treating injured birds rescued from oil spills to equine vets whose expertise was crucial to addressing the needs of neglected horses on nearby farms. This network of specialized professionals expanded our capacity to handle a broader range of cases, significantly enhancing the quality of care we provided. Before, a lack of specialized knowledge and resources often limited our treatment options. This collaboration meant that animals previously considered untreatable or requiring lengthy and costly referrals were receiving timely, effective care within our community.

Beyond veterinary professionals, I extended my collaboration to include animal behaviorists. Their insights proved invaluable in helping us rehabilitate animals suffering from trauma or behavioral issues. Animals that were initially fearful, aggressive, or withdrawn responded positively to behavior modification techniques, demonstrating the significant impact that such expertise can have on their behavior.

One remarkable example was a rescued dog, initially exhibiting severe aggression, who, after months of specialized training, transformed into a loving and gentle companion ready for adoption. This wasn't just about finding him a home; it was about restoring his capacity for trust and affection, demonstrating the power of collaborative care in healing not only physical wounds but also emotional scars.

The collaboration didn't stop with veterinary expertise. We engaged local animal shelters and rescue organizations. This collaborative effort helped us streamline rescue efforts by sharing resources, expertise, and coordinating admissions, minimizing stress for the animals involved. Instead of each organization operating in silos, we created a coordinated system that optimized the use of resources and ensured each animal received the level of care. This approach also extended to disaster relief efforts. When a wildfire threatened a nearby town, we could swiftly mobilize a combined force from different shelters and rescue organizations to evacuate hundreds of animals, demonstrating the effectiveness of our unified approach.

Engaging with local animal control agencies proved to be another vital collaborative endeavor. By fostering open communication and establishing shared protocols, we could improve response times and enhance the efficiency of animal rescue and rehabilitation efforts. For example, we collaborated with animal control to establish a rapid response system for cases involving neglected or abused animals, allowing for quicker intervention and improved animal welfare outcomes. This collaboration not only addressed immediate issues but also had a significant impact on educating the community about responsible pet ownership, leading to a decrease in neglect and abandonment cases in the long run.

The benefits of collaboration extended far beyond the direct care of animals. We have established partnerships with universities and veterinary schools, providing students with opportunities to gain hands-on experience while simultaneously enhancing our team's capabilities. This mutually beneficial relationship helped to train the next generation of animal welfare professionals, ensuring a continuous flow of skilled individuals committed to advancing the field. Students benefited from real-world experience, and our organization gained access to a vast pool of talent and innovative ideas. Such partnerships provided a fresh perspective, and a renewed energy, enhancing the dynamics within our organization.

We collaborated extensively with various community organizations, extending our reach far beyond the confines of our clinic. Partnerships with local schools enabled us to conduct educational programs for children, fostering compassion for animals and promoting responsible pet ownership from a young age. This long-term approach aimed at shaping a future generation of animal welfare advocates, reinforcing the broader community impact of our efforts. The collaboration also included engagement with local businesses, who generously provided supplies, volunteered their time, and helped us with fundraising initiatives. This highlighted the value of building positive relationships within the broader community, demonstrating that our work extended beyond the animals, changing community attitudes and behaviors.

One notable partnership involved a collaboration with a local rehabilitation center for individuals recovering from addiction. The center's residents volunteered at the clinic, fostering a sense of purpose and responsibility while contributing to the well-being of animals. This initiative showed the therapeutic benefits of animal interaction for individuals facing challenges, while simultaneously enhancing the workforce at our organization. This collaborative spirit extended to other community organizations, such as senior centers, where residents could interact with therapy animals, improving their overall well-being. These reciprocal partnerships showed the inextricable link between animal welfare and overall community well-being, dispelling the notion as an isolated endeavor.

Our successful collaboration also relied heavily on effective communication. Regular meetings with partner organizations, shared databases, and clear protocols ensured that information flowed seamlessly. We established a strong communication infrastructure built on mutual respect, transparency, and a shared vision. This transparency extended to keeping our community informed about our work and achievements, creating a shared sense of ownership and collective responsibility for animal welfare. This emphasis on communication was also vital in fostering trust among various stakeholders, ensuring that our collaborations remained strong and productive throughout the process.

These collaborations came with challenges. Differences in working styles and organizational cultures sometimes created hurdles, requiring careful attention and active listening. Compromise, mutual respect, and a shared vision were essential. Building trust and aligning goals took time, highlighting the patience and persistence needed for successful collaborative animal welfare.

Looking back, the collaborative efforts were not only instrumental in achieving significant improvements in animal welfare but also transformative in their impact on my personal growth as a veterinarian. Working alongside diverse professionals, engaging with various communities, and navigating the complexities of collaborative projects refined my leadership skills, improved my communication abilities, and strengthened my problem-solving skills. It was not merely about treating animals, but also about managing people, fostering relationships, and leading a team towards a common goal. This experience expanded my understanding of the broader societal context of animal welfare and enhanced my ability to address the challenges faced within the animal welfare sector effectively and efficiently. The journey of collaboration proved not only rewarding but also invaluable to my professional development, shaping me into a more effective leader and a more vigorous advocate for animal welfare. The impact resonates far beyond the number of animals helped; it has changed the way I approach my profession and contribute to the larger community.

Chapter 36: Defining Success: Personal Fulfillment

Moments of intense challenge and overwhelming joy have punctuated my journey as an aspiring veterinary student. The late nights hunched over textbooks, the pressure of examinations, the emotional toll of dealing with sick and injured animals–these were all significant hurdles. Yet, the sheer satisfaction of successfully diagnosing a complex condition, the palpable relief of seeing an animal recover, and the enduring bonds forged with colleagues and clients far outweighed any hardships. Looking back, however, I realize I don't define success solely by achievements, accolades, or even the tangible impact of my work. It's something far more profound and deeply personal.

True success, I understand, lies in the realm of personal fulfillment. It's about finding your passion, aligning your work with your values, and living a life filled with purpose and meaning. It's not about chasing external validation, social expectations, or accumulating wealth and prestige. These things can be byproducts of a successful life, but they aren't the essence of it. The external markers of success—the prestigious title, the financial rewards, the accolades received—pale compared to the inner satisfaction derived from a life lived authentically and with purpose.

My early years on the veterinary science competition team taught me the importance of hard work, dedication, and resilience. Those long hours of study and the single-minded chase for something just out of reach shaped my character and instilled in me a strong work ethic. These formed essential building blocks, but they only provided the foundation for a deeper understanding of success. The competition itself was exhilarating, and the adrenaline rush of performing under pressure was unforgettable. However, beyond the wins and losses, the competitions became a crucible in which I discovered a deep-seated passion for animal welfare and a burning desire to contribute meaningfully to the lives of animals.

The pursuit of my doctorate wasn't simply an academic exercise; it was a passionate journey fueled by a desire to enhance my capabilities and better serve the animal kingdom. The challenges were immense–the rigorous coursework, the demanding research, the pressure to succeed. Yet, with each hurdle overcome, and each milestone reached, it felt like a validation of my passion, a step closer to realizing my full potential. Not that the journey was easy; the emotional highs and lows of such a demanding program were often intense. There were moments of profound self-doubt, times when I questioned my abilities, and periods of exhaustion so profound they seemed insurmountable. However, what ultimately carried me through was the unwavering conviction that I was pursuing something significant, something that resonated with my core values.

My work in community outreach and fundraising further solidified this understanding. The tangible impact of improving animal welfare within my community was immensely gratifying. However, the precise measure of success was not solely the number of animals helped; this work highlighted the profound connection between personal fulfillment and meaningful contribution.

The collaborative nature of this work highlighted another critical aspect of success—the power of connection. Building relationships with colleagues, clients, and community members wasn't just about networking; it was about creating a support system, a shared vision, and a collective commitment to improving animal welfare. These connections weren't merely professional; they became deeply personal, enriching my life in ways that went far beyond the professional sphere. The friendships forged, the mutual respect developed, and the collective energy generated were all essential ingredients in achieving both professional and personal fulfillment.

Mentorship significantly contributed to my understanding of success. The guidance and support of my professors and mentors were invaluable, shaping my perspective and empowering me to overcome challenges I might not have otherwise faced. Their wisdom, encouragement, and belief in my potential were instrumental in my journey, demonstrating the profound impact of mentorship on achieving personal and professional goals. Their example underscored the importance of giving back, of guiding the next generation of veterinarians, and of fostering a culture of collaboration and mutual respect.

The challenges I faced in balancing my personal life with the demanding nature of my profession also played a significant role in defining my understanding of success. The struggle to find a healthy work-life balance and maintain meaningful relationships amidst the demands of my career was often intense. Yet, learning to prioritize my well-being, to nurture my relationships, and to create a life that was both fulfilling professionally and was an essential part of my journey. I realized I did not solely define professional achievement, but a harmonious integration of personal and professional life defined it.

Indeed, pursuiting success is not a linear path; it's a winding road with unexpected turns and unexpected detours. There will be moments of doubt, setbacks, and failures. But it is in these moments that we truly learn and grow. It's in overcoming challenges, navigating adversity, and learning from our mistakes that we discover our resilience, our capacity for growth, and our ultimate potential. This continuous process of self-discovery and adaptation is an integral part of what defines success. It's not about avoiding hardship, but about embracing it as an opportunity for growth and transformation.

We shouldn't consider the achievement of a particular destination the sole or definitive indicator of overall success, as many other factors contribute to a more holistic assessment. This is a journey of embracing challenges, surmounting obstacles through perseverance, and gaining valuable insights from any mistakes encountered along the way. Finally, the accurate measure of success lies not merely in the accomplishments one achieves, but in the profound personal growth and self-discovery that the journey entails.

The journey of veterinary medicine has been a testament to this truth—a path paved not only with professional accomplishments but also with profound personal growth, a deepened understanding, and an enduring sense of purpose. It's a journey I wouldn't trade for anything, and it's a testament to the fact that genuine success is far more than just reaching a destination; it's the enriching and strengthening experience of the journey itself.

Success is not a final destination, but an ongoing journey of growth and development. It's a journey of self-discovery, embracing one's true self, actively pursuing personal passions, and making a significant, positive impact on the world. The true essence of life lies in discovering joy during the journey, in embracing challenges as opportunities for growth, and in celebrating every accomplishment that marks our progress along the path. This journey is about discovering the equilibrium between your various commitments, carefully cultivating meaningful relationships, and consistently prioritizing your physical and mental well-being for a fulfilling life.

Ultimately, it's about finding a deep sense of purpose and fulfillment in everything you do. The pursuit of success, therefore, is not merely a professional endeavor but a deeply personal one – a reflection of who we are, what we value, and the impact we strive to make on the world. It's an ongoing evolution, a continuous journey of self-discovery and growth, where the destination is less important than the transformative process of getting there. And that, I believe, is the most accurate definition of success.

Chapter 37: The Value of Hard Work and Dedication

The demanding curriculum of veterinary school was a crucible, forging not just a veterinarian but a more resilient and determined individual. Those endless hours spent poring over textbooks, dissecting cadavers, and mastering complex anatomical structures weren't just about memorizing facts; they were about cultivating discipline, developing a profound understanding of the intricate workings of the animal body, and honing a meticulous approach to problem-solving. Each exam, each practical, each clinical rotation represented a challenge, a test of my knowledge, skills, and endurance. The weight of expectation, both self-imposed and external, was considerable. Failure wasn't an option; it wasn't even a consideration. The drive to succeed, to master the material, and to prove my capabilities to myself and others propelled me forward.

The pressure was immense, the workload daunting. Nights of little sleep, fueled by caffeine, left me overwhelmed by the sheer volume of information. Doubt crept in, whispering of inadequacy and failure, but I dismissed it. My commitment to animal welfare and passion for my path carried me through, enabling me to persevere and overcome each obstacle.

Beyond the academic rigors, the practical application of knowledge in the clinical setting was equally challenging. Working with sick and injured animals, witnessing their suffering, and bearing witness to the emotional toll on their owners was a profoundly formative experience. The intense emotions involved–the urgency of critical cases, the heartbreak of losing a beloved pet, the profound relief of successful treatment–shaped my approach to veterinary medicine, instilling in me a deep empathy and a dedication to providing the highest level of care. The emotional resilience I gained during these challenging clinical experiences was invaluable, shaping my ability to navigate the complex emotional landscape of veterinary practice.

The late nights in the hospital, the demanding surgeries, the countless hours spent comforting distraught owners–all these experiences, both positive and negative, shaped my understanding of veterinary medicine. They were far more than just professional experiences; they were profoundly personal ones that challenged me, strengthened me, and ultimately defined me as a veterinarian. The hard work wasn't just about acquiring knowledge and skills; it was about developing character, building resilience, and cultivating a deep sense of empathy and compassion.

My dedication extended beyond the classroom and the clinic. The countless hours spent volunteering at animal shelters, taking part in community outreach programs, and raising funds for animal welfare organizations showed my commitment to serving the larger animal community. This work wasn't just about contributing to the well-being of animals; it was about connecting with my community, sharing my passion, and making a tangible difference in the lives of others. The collaborative nature of this work underscored the power of community and the importance of collective action. These experiences broadened my perspective, fostering a deeper appreciation for the interconnectedness of animals, humans, and the environment.

Dedication to my studies and my community work wasn't always easy; it required significant sacrifices. There were social events I missed, personal relationships I neglected, and leisure activities I forwent. The pursuit of my goals required a significant personal commitment and a willingness to prioritize both my academic and professional aspirations. I never saw these sacrifices as hardships;

The value of mentorship and collaboration became increasingly apparent. The guidance of my professors, the support of my peers, and the collaborative spirit within the veterinary community were essential factors in my success. Learning from experienced professionals, collaborating on research projects, and engaging in meaningful discussions with colleagues enriched my understanding of veterinary medicine and fostered a deep sense of community and mutual respect.

The importance of effective time management, strategic planning, and unwavering self-discipline became increasingly crucial as I progressed through my studies. The ability to prioritize tasks, manage my time efficiently, and stay focused on my goals allowed me to navigate the demanding curriculum and maintain a healthy work-life balance. These skills, honed through years of dedicated effort and self-reflection, became indispensable assets not just in my academic pursuits but in all aspects of my life.

The pursuit of my doctorate was not just a journey of academic achievement; it was a transformative personal experience. It was a period of intense self-discovery, where I honed my skills, broadened my knowledge, and strengthened my resolve. The program's rigorous demands pushed me to overcome obstacles, challenge my boundaries, and ultimately emerge stronger and more capable. The perseverance I developed during this period became a cornerstone of my professional career and personal life.

My successes—awards, research projects, and challenging cases—were not merely personal triumphs; each accomplishment served as a powerful motivator, reinforcing my belief in my capabilities and fueling my drive to continue striving for excellence.

The failures I experienced—the research projects that didn't pan out, the exams on which I didn't score as well as I had hoped, and the cases where the outcome wasn't what I had wished for—were invaluable learning experiences. They taught me humility, resilience, and the importance of persistent effort. They challenged my assumptions, honed my critical thinking skills, and refined my approach to problem-solving. The lessons learned from these setbacks proved to just as valuable, if not more so, than the successes I celebrated.

My journey to earn my doctorate in veterinary medicine was an extensive and transformative process that went far beyond simply gaining the knowledge and skills; it was a profound personal and professional development experience. It was a period of profound personal growth and transformation, demanding not just intellectual brilliance but also exceptional emotional resilience, steadfast commitment, and a burning passion for the field I had pursued. Relentless hard work, unwavering dedication, and a steadfast belief in my capabilities to overcome challenges and achieve ambitious goals powerfully showed what I could accomplish. It stood as a clear testament to the profound fulfillment one can achieve by pursuing their passions and contributing significantly to society, highlighting the intrinsic rewards of meaningful work. More so than any career accomplishment, that defines success, a standard of achievement that surpasses all others. The journey, encompassing all of its inherent challenges and ultimate triumphs, proved to be its own unique and significant reward. The destination, while significant, was merely a culmination of a profound personal transformation, fueled by years of relentless effort, dedication, and unwavering belief in the power of hard work.

Chapter 38: The Importance of Support Systems, Relationships, and Mentorship

The culmination of my doctoral studies wasn't a solitary achievement; it was the crescendo of a symphony played by many. My dedication and perseverance formed the melody of my doctoral studies, while countless supportive relationships and invaluable guidance from mentors created the harmony. These weren't merely peripheral figures in my narrative; they were the pillars that sustained me through the most challenging times and propelled me towards success.

My parents, for instance, weren't simply providers of financial and emotional support. They were active participants in my journey. My mother, a teacher herself, understood the demands of rigorous academic pursuits. She fostered an environment that prioritized learning and intellectual curiosity, always encouraging me to pursue my passions, even when it meant making sacrifices. She would patiently help me organize my study schedule, prepare for exams, and even proofread my many research papers; her unwavering belief in my abilities was a constant source of strength and encouragement.

My father, a man of quiet strength and steadfast support, was always there with a listening ear and a comforting presence, particularly during those moments of self-doubt that inevitably surfaced during the arduous process. He instilled in me the importance of resilience and the conviction that setbacks were merely opportunities for growth. Their constant presence, quiet encouragement, and unwavering faith in my potential formed the bedrock of my academic achievements. They never pressured me, but empowered me to find my way and define my success on my terms. Beyond my family, my friendships served as essential lifelines amidst the academic tempest. These weren't superficial acquaintances; they were deep, meaningful connections forged in the crucible of shared experiences. My closest friends, fellow veterinary students, understood the relentless pressures and the overwhelming workload. They were my study buddies, my confidantes, and my emotional support system. Late into the night, we would often study together, our efforts sustained by a seemingly endless supply of coffee and the constant encouragement we provided to one another. In our shared experiences, we celebrated one another's triumphs and offered mutual support during setbacks, fostering a strong sense of camaraderie that went beyond the inherent competitiveness of our work environment. Throughout our personal challenges, we provided each other with unwavering support, creating a much-needed refuge from the rigorous demands of academic life.

They were my refuge, my escape, and the unwavering source of camaraderie that made the journey less arduous and infinitely more fulfilling. Their understanding ear was crucial in maintaining a sense of balance and perspective. The academic community also played a critical role. My professors, far from being distant figures of authority, were mentors and guides who invested in my growth as a veterinarian and as a person. Dr. Anya Sharma, my research advisor, played a crucial role in shaping my research trajectory. She provided invaluable guidance, challenging my assumptions and pushing me to think critically and creatively.

Her expertise, patience, and unwavering belief in my abilities helped to guide my research and ultimately shape my career aspirations. She taught me not only the technical aspects of research but also the importance of collaboration, perseverance, and the integrity of scientific inquiry. Dr. Ben Carter, my clinical supervisor, provided invaluable hands-on guidance, patiently teaching me the nuances of clinical practice, guiding my judgment in crucial decision-making, and instilling in me a deep sense of empathy and compassion for my patients and their owners. His mentorship was a blend of rigorous instruction and encouraging support, forming the cornerstone of my development as a veterinarian. The consistent guidance and unwavering support of these mentors were transformative, shaping not only my professional skills but also my personal growth. They showed me that success was less about individual achievement and more about collective growth and mutual support.

Beyond the formal mentoring relationships, the informal interactions within the veterinary community also proved immensely valuable. Attending conferences, taking part in workshops, and engaging in informal discussions with experienced professionals enriched my understanding of the field and broadened my network of peers. These interactions built camaraderie and collective responsibility, creating a supportive environment where people freely shared knowledge and experience, thus fostering mutual growth and support.

The importance of mentorship extended beyond the formal academic realm. My involvement in community outreach programs and volunteer work at animal shelters exposed me to a diverse range of individuals, each with their own expertise and experiences. Veterinary technicians, animal shelter staff, volunteers, and community members shared their perspectives, insights, and invaluable experiences, expanding my understanding of animal welfare, community engagement, and the broader impact of veterinary medicine. These interactions were enriching, broadening my knowledge of the field beyond the confines of the academic setting and providing a holistic perspective on animal care and community engagement.

The relationships were not simply helpful; they were essential for success, providing both emotional and practical sustenance. Throughout my journey, they offered the emotional support, intellectual inspiration, and practical advice that were essential for overcoming obstacles and appreciating achievements. The constant reminder of their presence underscored the truth that success is not a solitary pursuit, but a collaborative endeavor, a rich tapestry composed of the interwoven threads representing countless connections between individuals.

Reflecting upon my journey, it's clear that the strength of my support system was directly proportional to the magnitude of my achievements. The challenges encountered were not only overcome through personal dedication but also through the unwavering support, guidance, and collaboration of those who believed in my potential. The empathy and understanding I received were as crucial as any textbook or clinical experience could be. This realization profoundly affected my perspective, not just on success, but also on human connection and the importance of building meaningful relationships. It underscored the understanding that success is not solely the product of individual effort but a collaborative achievement, facilitated and fostered by the collective support and guidance of a network of individuals. This understanding became a cornerstone of my professional philosophy, shaping my interactions with colleagues, students, and clients and reinforcing the importance of community and collaboration in all aspects of life.

Earning my doctorate was more than an academic pursuit; it was an exploration of human connection, resilience, and the power of support. The relationships I built with family, friends, mentors, and colleagues were essential, providing the emotional strength, intellectual stimulation, and guidance needed to navigate my studies and achieve my goals.

Chapter 39: Overcoming Challenges: Resilience and Adaptability

The path to my doctorate wasn't a straight line; it was a winding road punctuated by unexpected detours and steep inclines. There were moments of profound self-doubt, periods of intense pressure, and instances where failure felt like an inescapable reality. Yet, it was in navigating these challenges that I discovered the actual depth of my resilience and the importance of adaptability.

One particularly challenging hurdle arose during my research. My initial hypothesis, meticulously crafted and rigorously tested, yielded unexpected and, frankly, disappointing results. The data didn't support my assumptions. The initial reaction was devastating. Months of painstaking work, countless hours in the lab, and the weight of expectation felt like they had all been for naught.

The temptation to abandon the project entirely was almost overwhelming. Self-doubt gnawed at me, whispering insidious suggestions of inadequacy and failure. The thought of starting over, revising my approach, and potentially facing further setbacks was daunting.

But the support system I had cultivated over the years proved invaluable during this crisis. Dr. Sharma, my mentor, didn't offer platitudes or dismiss my feelings. Instead, she acknowledged the significance of the setback and provided a space for me to process my emotions. Her calm guidance helped me reframe the situation. She reminded me that scientific research is inherently iterative, a process of hypothesis, testing, refinement, and sometimes even discarding initial assumptions. The failure, she emphasized, wasn't a reflection of my capabilities, but an integral part of the learning process. This perspective shift was crucial; it transformed a crushing defeat into a valuable opportunity for growth.

With renewed resolve, I meticulously reviewed my method, scrutinizing every step, every assumption, and every potential source of error. I sought feedback from colleagues, engaging in open and honest discussions about my findings. Their input, their insights, and their perspectives shed light on aspects I hadn't previously considered. Through this collaborative effort, we identified flaws in my experimental design and refined my approach. It wasn't just about fixing the technicalities; it was about cultivating a more robust and nuanced understanding of the research question itself. The revised study, with its improved method, ultimately yielded interesting results that exceeded my initial expectations. The triumph wasn't simply about producing publishable data; it was about demonstrating the power of perseverance, the importance of critical self-reflection, and the invaluable role of collaboration in overcoming scientific setbacks.

Beyond research, clinical practice presented its own unique set of challenges. The emotional toll of dealing with sick and injured animals, the pressure of making critical decisions under time constraints, and the occasional confrontations with distraught pet owners tested my emotional resilience and problem-solving abilities. One particular instance stands out vividly. A young, severely injured dog arrived in the emergency room, critically ill and unresponsive. Despite our best efforts, the situation quickly deteriorated. The owner, understandably distraught, was desperate for a miracle. The weight of responsibility, the potential for failure, and the emotional impact of witnessing the animal's suffering were immense. I felt the pressure intensely, a knot of anxiety tightening in my chest.

However, rather than succumbing to panic, I consciously applied the problem-solving skills I'd honed throughout my education. I methodically assessed the dog's condition, analyzing the symptoms and considering various diagnostic possibilities. I consulted with experienced colleagues, seeking their expertise and input. The collaborative approach, characterized by calmness and systematic, along with focusing collectively on the animal's well-being, eased the pressure and fostered a sense of shared responsibility. While the outcome wasn't ultimately a complete victory–the dog's injuries were too severe–the experience taught me the immense value of teamwork, the importance of clear communication, and the necessity of maintaining composure under pressure. It highlighted the necessity of emotional resilience, not only for myself but also for the veterinary team, enabling us to provide the best possible care, even in the face of a difficult prognosis.

Adaptability proved to be another crucial factor in overcoming these challenges. The veterinary field is constantly developing, with advancements in technology, research, and treatment protocols. Staying current required a commitment to continuous learning, a willingness to embrace new techniques, and an openness to adapting my approach in response to changing circumstances. This was true during my clinical rotations, where I encountered a wide range of cases, each demanding a unique approach. I had to be flexible, resourceful, and able to adapt quickly to various situations, diagnostic tools, and therapeutic interventions. This involved constant learning and updating my knowledge base. I attended workshops, conferences, and seminars, where I devoured new research findings and explored innovative treatment strategies. By continuously learning and adapting, I was always ready for new challenges.

Adaptability extended beyond technical skills. It encompassed my approach to communication, my interactions with clients, and my ability to navigate the emotional complexities of veterinary medicine. Learning to communicate effectively with clients facing hard decisions, managing their expectations, and offering emotional support during challenging situations was paramount. This involved mastering empathy, active listening, and clear and concise communication. Building trust and rapport with clients, understanding their perspectives, and tailoring my approach to their individual needs were key to providing adequate care. This adaptation extended beyond clinical scenarios; it also influenced my approach to personal challenges, including adapting to different learning styles, adjusting my study habits to accommodate various workloads, and cultivating self-care practices to manage stress levels.

The pursuit of my doctorate was not merely an academic endeavor; it was a crucible that forged within me resilience, adaptability, and a profound understanding of the importance of support and collaboration. The setbacks I faced, while challenging and disheartening, ultimately contributed to my growth, sharpening my skills, deepening my understanding, and strengthening my resolve. Each obstacle overcome served as a stepping stone, propelling me forward on my journey, forging within me a strength and adaptability that I never knew I possessed. The culmination of my trip wasn't simply the acquisition of a degree; it was the development of a resilient spirit, the cultivation of problem-solving skills, and a deep understanding of the transformative power of human connection. This understanding informs my practice as a veterinarian, underpinning my commitment to providing compassionate care and empathetic support to both animals and their owners. My journey underscores that genuine success is not merely achieving a goal, but the personal growth that comes from navigating the challenges that lie along the way.

Chapter 40: Balancing Personal Life and Career: Maintaining Wellbeing

The relentless pursuit of my doctorate had consumed me — a whirlwind of late nights in the library, early mornings in the lab, and endless clinical rotations. The exhilaration of discovery and the satisfaction of mastering complex procedures were undeniable. Still, a creeping fatigue settled over me, a subtle yet persistent weariness that hinted at a more profound imbalance. The line between my professional ambition and my well-being had blurred, threatening to unravel the very fabric of my life.

It wasn't a sudden collapse; it was a gradual erosion, a slow bleed of energy and enthusiasm. The vibrant passion that had fueled my journey dimmed, replaced by a sense of relentless pressure and a gnawing anxiety. Sleep became a luxury, meals an afterthought, and social interactions a distant memory. My body, once a resilient engine driving me forward, sent out distress signals–persistent headaches, a weakened immune system, and an overwhelming sense of exhaustion that lingered even after hours of rest. My deteriorating health served as a wake-up call, starkly reminding me that even the most ambitious goals shouldn't come at the expense of my health and happiness.

The realization struck me with the force of a physical blow. I had been so focused on achieving external validation–the prestigious doctorate, recognition from colleagues, and the respect of my mentors–that I had neglected the most fundamental aspect of success: my well-being. The pursuit of excellence had become self-neglect, a relentless striving that left me depleted and vulnerable. It was a painful lesson, a hard-earned truth that resonated deep within me. I needed to recalibrate and redefine success not just as the achievement of a goal, but as a holistic state of well-being encompassing my physical, mental, and emotional health.

This realization sparked a shift in perspective. I began prioritizing self-care, integrating practices that nurtured my physical and mental well-being. Regular exercise, once neglected, became essential, not just for fitness, but to clear my mind, release stress, and reconnect with my body. Even short walks in nature or quiet city streets offered a welcome respite from the relentless demands of my studies.

The simple act of breathing fresh air, feeling the sun on my skin, and observing the quiet rhythm of nature proved incredibly restorative.

Sleep became a priority, a sacred time for rest and rejuvenation. I established a consistent sleep schedule, ensuring I got enough restful sleep each night. I created a calming bedtime routine that minimizes screen time and encourages engaging in relaxing activities, such as reading or listening to soothing music. These changes significantly improved the quality of my sleep and my overall well-being. The impact on my concentration, my mood, and my ability to cope with stress was immediate and remarkable.

Nutrition, once an afterthought, also received a significant upgrade.

I transitioned to a more balanced diet, focusing on whole foods, fruits, and vegetables. I reduced my intake of processed foods, sugary drinks, and excessive caffeine, which had contributed to my energy crashes and mood swings. The shift in my diet provided a sustained energy boost, reducing fatigue and enhancing my overall mental clarity.

Mindfulness practices also played a significant role in my journey toward a balanced life. I incorporated meditation into my daily routine, even if it was only for a few minutes each day. The practice of focusing on my breath, observing my thoughts without judgment, and cultivating a sense of present moment awareness proved incredibly effective in managing stress and anxiety. It provided a space for introspection, allowing me to process my emotions and regain a sense of calm amid the chaos.

Beyond these individual practices, I also made conscious efforts to integrate social interactions into my life. I reconnected with old friends, made new ones, and cultivated meaningful relationships that brought joy and support into my life. These connections served as anchors, grounding me during challenging times and providing a much-needed sense of belonging. Sharing my experiences, struggles, and triumphs with others proved incredibly cathartic, reminding me I wasn't alone on my journey.

The achievement of a healthy work-life balance wasn't a single monumental event, but a gradual process of adjustments and compromises. To accomplish this, I had to set boundaries, learn to decline commitments that sapped my energy, and, where possible, successfully delegate tasks to others. Actively seeking support involved reaching out to my friends, family, and mentors, embracing vulnerability, and openly asking for help whenever challenges arose.

The transformation wasn't solely about external changes; it was equally about cultivating inner resilience — a self-compassionate approach that allowed me to acknowledge my limitations without self-criticism. I learned to celebrate minor victories, to recognize and appreciate my accomplishments, and to forgive myself for setbacks.

This self-acceptance was crucial in building self-esteem and fostering a sense of self-worth that extended beyond my professional achievements.

The impact of this intentional focus on well-being was profound. My energy levels soared, my focus sharpened, and my overall sense of well-being significantly improved. The relentless pressure and gnawing anxiety that had once dominated my life dissipated, replaced by a sense of calm and inner peace. I discovered that a balanced life wasn't a compromise; it was a catalyst for enhanced productivity and creativity. My work became more fulfilling, my relationships more meaningful, and my overall sense of purpose deepened.

This journey of self-discovery and personal growth wasn't just about maintaining a healthy work-life balance; it was about cultivating a life that was truly aligned with my values and aspirations. It was about understanding that success wasn't a destination, but a continuous process of growth — a journey of self-discovery, resilience, and the unwavering pursuit of well-being. The doctorate was a significant accomplishment, but the real reward was the strength, resilience, and self-awareness I had cultivated along the way. This newfound balance wasn't just a temporary fix; it became the bedrock upon which I built a fulfilling and sustainable life, one where my professional ambitions and personal well-being existed in harmonious collaboration. The path to success, I learned, wasn't a solitary climb; it was a journey enriched by self-care, supportive relationships, and a deep understanding of the interconnectedness between inner peace and outward achievement. It was a journey of continuous learning, adaptation, and an unwavering commitment to nurturing both my professional aspirations and my well-being–a delicate balance that ultimately defined my genuine success.

And that, more than any academic achievement, was the greatest reward of all.

Chapter 41: Morning Routine: Preparing for the Day

The alarm's gentle chime, a melody far removed from the jarring shriek of my earlier, more stressed days, pulled me from sleep. Sunlight, muted by the sheer curtains, painted soft stripes across my bedroom floor. This wasn't the frantic scramble of previous mornings, fueled by caffeine and anxiety. This was a deliberate, mindful awakening. I lay for a few moments, savoring the quiet stillness before the day's demands encroached. This pause, this conscious choice to start the day slowly, was a crucial element in the newfound equilibrium I'd carefully cultivated.

Each morning, I followed a ritualistic routine, a sacred and personal time set aside for self-care and reflection. Rather than strictly following a timetable, I prioritized a more adaptable and fluid approach to my schedule. To start my day, I began with a gentle stretching routine, a carefully planned sequence of movements designed to awaken my muscles and ease any lingering stiffness from the physically demanding activities of the previous day. I focused intently on each stretch and breath, using deliberate, slow, and careful movements to gradually dissipate the muscle tension with every measured action. Previously neglected and mistreated, my physical form had undergone a transformation; it was now a source of strength and resilience, a vessel worthy of my utmost respect and nurturing care.

Following the stretching, I would head to the kitchen, where the aroma of freshly brewed coffee already hung in the air. I'd prepared the coffee the previous evening, a minor act of self-care that saved precious minutes in the morning. This wasn't just any coffee; it was a carefully selected blend, a rich and aromatic indulgence that felt like a small luxury, a reward for the day ahead. As I sipped my coffee, I would sit by the window, observing the city slowly awakening, watching the sun paint the sky with hues of gold and rose. This quiet contemplation, this mindful appreciation of the ordinary, was a vital part of my morning routine. It allowed me to ground myself, to center myself before the day's challenges began.

Breakfast was another mindful affair, a conscious choice to nourish my body with healthy, wholesome food. This wasn't a rushed meal; it was an opportunity to savor the flavors, appreciate the textures, and nourish myself from the inside out. My plate was a vibrant tapestry of colors, a balanced selection of fruits, vegetables, and whole grains, far removed from the processed foods and sugary snacks that had once fueled my relentless pursuit of my doctorate. The food was more than just sustenance; it was an act of self-love, a declaration that I valued my body and its well-being.

After breakfast, I'd dedicate time to mindfulness practices, usually a guided meditation or a session of deep breathing exercises. These practices weren't just about relaxation; they were about cultivating a sense of inner peace and self-awareness. Focusing on my breath and gently observing my thoughts and emotions helped me manage stress and anxiety, preparing me for the emotional challenges of the workday. It was a time for introspection, for tuning into my inner self, and for reminding myself of my values and aspirations. These moments of stillness, even amid a busy day, were crucial in maintaining my mental and emotional well-being.

After my mindfulness practice, I'd review my schedule for the day, mentally preparing myself for the tasks ahead. This wasn't a mere glance at my calendar; it was a thoughtful review, a conscious engagement with my day's responsibilities. I would prioritize my tasks, setting realistic goals and allocating specific time slots for each activity. This careful planning minimized the feeling of being overwhelmed and helped me to maintain a sense of control over my schedule. The sense of preparedness empowered me, fueling my confidence and making me more receptive to the challenges and opportunities that the day presented.

The preparation for my workday extended beyond the physical and mental aspects of my work. I paid close attention to my attire. Choosing my outfit wasn't a hurried decision; it was a mindful act, reflecting my commitment to presenting myself professionally and feeling confident in my appearance. Dressing professionally wasn't just about adhering to a dress code; it was about projecting an image of competence and self-respect—this attention to detail, this care in my appearance, extended to my work environment as well. I organized and tidied my workspace, creating a calm and productive atmosphere.

This morning routine wasn't a rigid schedule; it was a flexible framework designed to adapt to the day's demands. Some days, I might spend a little more time meditating, or perhaps I'd enjoy a longer walk in the park. On other days, I might need to allocate more time to reviewing my schedule or preparing for a specific case. The key was flexibility — the ability to adjust my routine to accommodate the ebb and flow of life.

The most important aspect of this carefully constructed routine was the sense of intentionality it fostered. Each activity, each moment, was a conscious choice, a deliberate act of self-care. It wasn't about achieving perfection, and it was about creating a framework that supported my physical, mental, and emotional well-being. It was about starting the day with a sense of calm, clarity, and purpose, setting the tone for a productive and fulfilling day ahead. This foundation, this mindful approach to my morning routine, transformed my workday from a source of relentless pressure and anxiety into an opportunity for meaningful work and personal growth. It wasn't just about preparing for the day; it was about preparing myself for the challenges and opportunities that lay ahead, ensuring I had the physical and mental resources to navigate them with grace and resilience.

The transition from my carefully crafted morning routine to the demands of my day at the clinic was seamless, a testament to the effectiveness of my newfound self-care practices. My work, which once caused overwhelming stress, now feels manageable and even fulfilling. The calm I cultivated in the morning resonated throughout my day, helping me approach challenges with a sense of clarity and focus. The meticulous attention to detail I applied to my morning routine translated into my patient care, enhancing my ability to diagnose illnesses and deliver effective treatments.

My days were a vibrant mix of routine and unexpected challenges–the comforting predictability of scheduled appointments juxtaposed with the adrenaline rush of emergency cases. I navigated this duality with newfound composure, my resilience honed by the self-care practices that had become an integral part of my life. The day's challenges, once daunting, were now viewed as opportunities for growth and learning. The success of my interventions–a kitten's recovery from pneumonia, a dog's relief from chronic pain, the comforting of a distressed pet owner–reinforced my commitment to my profession and brought a deep sense of fulfillment that extended far beyond the achievement of my doctorate.

Evenings, too, were a deliberate departure from the past's relentless work ethic. My evenings were no longer consumed by endless studies or the anxiety of impending deadlines. Instead, I dedicated my evenings to unwinding and nurturing my relationships. I made time for dinner with friends, engaging in conversations that nourished my soul and reminded me of the support network that had been instrumental in my success. There were quiet evenings spent reading or listening to music, activities designed to soothe my mind and prepare me for the next day's adventures. I had consciously cultivated a balance, creating a life where professional ambition and personal well-being thrived side by side — a harmonious coexistence that enriched my life immeasurably. The relentless pursuit of my doctorate had once defined me.

Still, it was now merely one chapter in a much larger, more fulfilling narrative—a story of continuous growth, resilience, and the unwavering pursuit of a life lived in balance. The achievement of my doctorate was undoubtedly a significant milestone. Still, the valid reward, the ultimate achievement, was the life I had cultivated–a life of purpose, balance, and enduring joy.

Chapter 42: Clinical Cases, Diagnoses, and Treatments

The first patient of the day was a six-year-old Bloodhound named Gus, who presented with lethargy and a persistent cough. His owner, a worried young woman named Karen, recounted how Gus had been gradually losing energy over the past couple of weeks, and the cough had started just a few days ago. The initial physical examination revealed a slightly elevated respiratory rate and some crackling sounds in his lungs. I suspected kennel cough, a highly contagious respiratory infection common among dogs, but I ordered a chest X-ray to rule out pneumonia or other more serious conditions. The X-ray confirmed my suspicions; there was evidence of mild inflammation in the airways, consistent with a viral infection. I prescribed Gus a course of antibiotics to combat any secondary bacterial infections, along with a cough suppressant to ease his discomfort. I also advised Sarah on supportive care, including keeping Gus warm, providing plenty of fluids, and ensuring he received adequate rest.

My next patient was a three-month-old Persian kitten named Snowball, suffering from severe diarrhea. Her owner, a senior gentleman named Mr. Henderson, was visibly distressed. Snowball was lethargic and dehydrated, showing signs of significant weight loss. I immediately suspected feline infectious enteritis, a potentially fatal viral infection. I performed a fecal examination to confirm the diagnosis and ruled out other causes of diarrhea, such as parasites or dietary indiscretions. The fecal test confirmed the virus. Treatment focused on supportive care, including intravenous fluid therapy to correct dehydration and electrolyte imbalances, and anti-emetic medication to control vomiting. I also prescribed antibiotics to prevent secondary bacterial infections. Mr. Henderson was understandably anxious, but I reassured him that with prompt treatment, Snowball had a good chance of recovery. Daily updates and close monitoring of her condition were vital. I advised Mr. Henderson on strict hygiene measures to prevent the spread of the infection and provided him with detailed instructions on administering the medications.

Later that morning, a frantic call shattered the calm. A car had struck a young Labrador, Max, and his owner, a teenage girl named Emily, was sobbing uncontrollably. I instructed her to bring him to the clinic immediately. Upon arrival, Max was unconscious, with shallow breathing and a weak pulse. We acted swiftly, securing his airway, administering oxygen, and starting intravenous fluids to support his blood pressure. A thorough examination revealed a fractured femur and internal bleeding. I immediately referred Max to a nearby veterinary hospital equipped for emergency surgery. I accompanied Emily and Max, providing support and explaining the procedures to them. The veterinary surgeons at the hospital were excellent, and the surgery was a success. The veterinary surgeons repaired the fracture and controlled the internal bleeding. After the surgery, the veterinary staff placed Max in intensive care for close observation. Emily's anxiety was palpable, but I reassured her that Max was in the best possible hands.

The afternoon brought a more routine case. We brought in a ten-year-old domestic shorthair cat named Mittens for a routine check-up and vaccinations. Her owner, a middle-aged woman named Mrs. Davis, was meticulous about Mittens's health and always ensured she was up-to-date on her vaccinations. The examination revealed Mittens to be in excellent health. I administered her vaccinations and discussed senior cat care with Mrs. Davis, highlighting the importance of regular check-ups to detect and address any age-related changes.

The last patient of the day was a five-year-old German Shepherd named Quendi, presented with lameness in her right hind leg. Her owner, a middle-aged man named Mr. Johnson, explained that Quendi had been limping for the past few days, and the limping was worsening. The physical examination revealed swelling and tenderness in her right knee. I suspected a cruciate ligament rupture, a common injury in dogs. I ordered X-rays to confirm the diagnosis and rule out other causes of lameness, such as bone fractures or infections. The X-rays confirmed a partial rupture of the anterior cruciate ligament. I discussed various treatment options with Mr. Johnson, including conservative management with pain medication and physical therapy, or surgical repair.

Given Quendi's age and health, we chose a conservative approach. I prescribed pain and anti-inflammatory medications and referred Mr. Johnson to a canine rehabilitation specialist for physical therapy. From routine vaccinations to life-threatening emergencies, each case presented unique challenges, demanding careful consideration and meticulous attention to detail.

We tailored each treatment plan to the individual needs of the patient, considering their age, breed, overall health, and the severity of their condition. The day was a whirlwind of activity, a blend of the mundane and the dramatic, reflecting the ever-changing nature of veterinary practice. But throughout it all, the common thread was a deep commitment to compassionate care, a dedication to easing suffering, and the immense satisfaction of witnessing the healing process and the joy of reuniting patients with their relieved and grateful owners.

One of the most rewarding aspects of my work was the opportunity to forge strong bonds with my patients and their owners. Building trust and rapport is crucial, not only for effective diagnosis and treatment but also for providing emotional support during times of stress and anxiety. The bond between a pet and its owner is often profound, and I strived to approach each interaction with empathy and understanding, offering not only veterinary expertise but also a listening ear and a compassionate heart. This patient-centered approach extended beyond the clinical examination and treatment; it encompassed educating owners on preventive care, providing dietary advice, and guiding them through the recovery process.

The job takes a significant emotional toll. Witnessing animal suffering, dealing with grieving owners, and facing the limitations of veterinary medicine can be emotionally draining. However, the rewards far outweigh the challenges. The joy of seeing a sick animal recover, the gratitude of an anxious owner, and the knowledge that I am making a difference in the lives of both animals and their people sustain me. These moments of connection, these minor victories, are what make the work meaningful, challenging, and incredibly fulfilling. It's a career that demands dedication, resilience, and compassion, yet it's also a career that's richly rewarding.

The journey to becoming a veterinarian wasn't merely about achieving a professional goal; it was about discovering a profound connection to the natural world, a calling to serve, and a lifelong commitment to the welfare of animals. And this, I felt, was more rewarding than any professional achievement. The satisfaction wasn't just in the successful diagnoses and treatments, but in the subtle moments–a grateful lick from a recovering pup, the relieved sigh of a worried owner, the quiet understanding that passes between animal and doctor. These are the proper measures of a fulfilling life in veterinary medicine. The day ended, not with exhaustion, but with a quiet satisfaction. The cases, though diverse, all shared a common thread–the opportunity to make a positive impact, to heal, and to offer comfort. And that, I realized, was the greatest reward of all.

Chapter 43: Interactions with Clients: Communication and Compassion

The next day began with a flurry of phone calls. A frantic voice described a cat tangled in fishing line, a choked whimper punctuating the urgency. Another was a concerned owner reporting a dog that had ingested an unknown substance, exhibiting mild distress symptoms. These calls, though routine in their frequency, highlighted the diverse spectrum of challenges faced daily. Effective communication, I realized, wasn't merely about conveying medical information; it was about providing reassurance, managing expectations, and fostering a collaborative relationship built on mutual trust and respect.

Mr. Mittens, a cat, and Mr. Fitzwilliam, a visibly agitated retired schoolteacher, were in the same place. Mittens, his beloved Siamese cat, got hopelessly tangled in the line, causing minor injuries and significant distress. I soothed his worries by calmly outlining our safe line removal process to prevent further injury. Throughout, I used soothing words to keep Mittens calm. This small display of empathy, as I discovered, lessened my worry over Mittens and Mr. Fitzwilliam. Mr. Fitzwilliam expressed immense gratitude following the successful removal. His relief stemmed from Mittens's physical recovery and the emotional support I provided. He spent several minutes detailing Mittens's history, emphasizing his care for her and explaining his worries, showing that these interactions balanced treatment with listening and understanding.

The dog, a boisterous Labrador named Buddy, presented a distinct challenge. His owner, a young mother called Lisa, was worried, but the initial examination revealed no immediately apparent distress. Buddy was active, albeit slightly listless. Lisa's thorough description of his behavior, including the specific time he ingested the unknown substance, was vital in determining the course of action. Here, good communication extended beyond obtaining facts; it involved educating Lisa about the potential risks associated with unknown substances and guiding her through a process of closely monitoring Buddy. I advised her on the signs to watch for–vomiting, lethargy, diarrhea–and stressed the importance of contacting me immediately if any concerning symptoms developed. I also gave her my number for after-hours communication, ensuring she felt supported and empowered to make informed decisions about Buddy's care.

These two cases, though vastly different, shared a common thread: the importance of building rapport with clients. The art of veterinary medicine, I realized, is not solely about treating animals, but about building relationships with their owners, nurturing their emotional needs, and understanding their worries and concerns. Often, their anxiety is as palpable as the animal's physical distress. It's about listening actively, providing simple explanations in terms they understand, and answering their questions thoroughly, even if it involves repeating myself multiple times.

Later that afternoon, a senior couple arrived with their aging Beagle, a sweet-natured dog named Pepper. Pepper was suffering from age-related decline, his joints stiff and his movements labored. The owners, Mr. and Mrs. Anderson, were understandably distraught. They'd shared twenty years with Pepper, and the prospect of his declining health weighed heavily on them. This situation highlighted the emotional complexity of veterinary practice. Besides offering medical advice and treatment options for Pepper's arthritis, I spent a considerable amount of time listening to the Andersons recount their memories of Pepper–his playful puppyhood, his unwavering loyalty, the joy he brought into their lives. I validated their feelings, acknowledged their grief, and offered practical advice on managing Pepper's pain and maintaining his quality of life. This consultation extended well beyond the typical veterinary examination; it became a shared experience of navigating the delicate transition of an aging beloved companion.

Throughout the day, I reflected on the diverse array of emotions that flowed through the clinic—relief, anxiety, gratitude, and grief. These were not just my emotions, but those of my clients, intertwined with their beloved animals. The ability to connect with them on these emotional levels was, I recognized, just as crucial as my medical expertise. A simple gesture, a comforting touch, a reassuring word could often ease anxiety more effectively than any medication.

The act of listening attentively, validating their feelings, and offering a space for them to share their worries and concerns was a critical component of my role.

One particularly memorable interaction involved a young girl named Lily, who had brought in her hamster, a fluffy creature named Pip. Pip had suffered a minor injury — a minor cut on his paw. While the injury itself was minor, Lily's distress was palpable. She was overwhelmed with worry and felt utterly responsible for Pip's well-being. I treated the wound, but I also took the time to reassure Lily, explaining the treatment process clearly and simply, and answering her questions with patience and empathy. I allowed her to help me, showing her how to apply a small bandage. This minor act of inclusion empowered her, shifting her focus from helplessness to involvement. Lily left the clinic with a lighter heart, feeling that, both her hamster and she had received care and support.

The day concluded with a profound sense of fulfillment. The satisfaction wasn't simply from successfully diagnosing and treating animals, but from the connections forged with their owners. These interactions were a testament to the holistic nature of veterinary medicine—it's a vocation that intertwines medical expertise with emotional intelligence, compassion, and a deep understanding of the bond between humans and animals. It requires empathy, patience, clear communication, and a profound knowledge of the distinct dynamics of each human-animal relationship. The challenges are considerable, the emotional toll significant, yet the rewards-the gratitude in the eyes of a relieved owner, the gentle nuzzle from a recovering pet, the knowledge that I have played a role in easing suffering and strengthening bonds—are immeasurable. It's a career that continues to challenge, teach, and deeply fulfill me. The journey continues, not just as a veterinarian, but also as a compassionate caregiver and a listener — a role that is as vital as any medical intervention. The human element, I realized, is the cornerstone of this fulfilling and gratifying profession.

Chapter 44: Teamwork and Collaboration: Working with Colleagues

The next morning dawned bright and crisp, promising a day as busy as the last. However, this day presented a unique challenge: a complex case that required the coordinated expertise of our entire team. Someone brought in Gus, a young, spirited golden retriever, with a severe lameness in his hind leg. Initial examinations by Dr. Ramirez, our resident orthopedic specialist, revealed a suspected cruciate ligament rupture–a complex injury requiring surgical intervention. This wasn't a solo operation; this was a collaborative effort, a symphony of veterinary skills orchestrated to give Gus the best possible chance of recovery.

Dr. Ramirez, with his years of experience in orthopedic surgery, took the lead, meticulously outlining the surgical plan. His expertise was crucial, but his role extended beyond the operating table. He briefed the rest of us–myself, Dr. Chen, our resident anesthesiologist, and Sarah, our experienced veterinary technician–on the specifics of Gus's condition, the surgical procedure, and the potential complications. His clear and concise communication ensured everyone agreed, fostering a sense of shared responsibility and collective commitment to Gus's well-being.

Dr. Chen, a master of her craft, played a critical role in ensuring Gus's safety during the surgery. Her pre-operative assessment was thorough, meticulously evaluating Gus's heart and respiratory function, ensuring he was fit for anesthesia. Throughout the surgery, her vigilance was unwavering, continuously monitored Gus's vital signs, making subtle adjustments to maintain optimal anesthetic levels. Her calm efficiency was reassuring, a silent testament to her deep expertise. She didn't simply monitor; she actively took part, providing invaluable feedback to Dr. Ramirez, alerting him to any subtle changes in Gus's condition. Their seamless communication during the procedure was a testament to years of working together, a comfortable rhythm of understanding that minimized potential risks and maximized Gus's safety.

Sarah, our veterinary technician, proved to be an indispensable part of the team. Her experience in surgical preparation, her deft hand in assisting Dr. Ramirez, and her calm presence in the operating room were instrumental to the success of the surgery. She meticulously prepared the surgical instruments, maintaining sterile conditions, anticipating Dr. Ramirez's needs, and passing him the right tool at precisely the right moment. Her attentiveness wasn't just about efficiency; it showed a profound understanding of the surgical process, a recognition that even the most minor details could have significant consequences. She monitored Gus's post-operative recovery, carefully observing for any signs of complications. Her meticulous record-keeping documented all aspects of Gus's care thoroughly, contributing to a comprehensive understanding of his treatment.

The success of the surgery wasn't merely a reflection of individual skills, but a testament to the power of collaboration. Each team member brought a unique set of expertise, working together seamlessly to overcome the challenges posed by Gus's injury. The surgery presented many challenges and required a high level of skill and precision from the surgical team. Because of the smooth flow of information, a shared understanding of the surgical plan, and seamless task coordination, the surgery had a successful outcome; this success would have been unattainable, or at the very least significantly more difficult, had each member of the team worked independently.

The post-operative care was equally demanding, requiring consistent monitoring and meticulous attention to detail. Gus's recovery was slow but steady, each slight improvement a cause for celebration among the team. The daily check-ups involved not just assessing physical recovery but monitoring Gus's emotional well-being. We all contributed to his comfort and well-being, ensuring that he felt secure and cared for, which facilitated a better recovery. This shared responsibility extended beyond the medical aspects; it encompassed a profound sense of collective ownership in Gus's recovery.

This experience reinforced the profound understanding that veterinary medicine is not a solo endeavor. It thrives on teamwork, collaboration, and a shared commitment to providing the best possible care. Our success wasn't merely the result of individual brilliance; it was a collective triumph, a testament to the power of working together, supporting each other, and drawing strength from the diverse expertise within our team.

Another instance highlighted the importance of cross-departmental collaboration. A frantic call came in about a flock of sheep displaying unusual lethargy and loss of appetite. Dr. They immediately dispatched Evans, our large animal specialist, to the farm. However, as the initial assessment unfolded, the problem went beyond simple illness. The sheep showed signs of digestive problems, and we suspected a toxin had contaminated their feed. Dr. Evans requested help from Dr. Lee, our expert in toxicology. Their collaboration began immediately. Dr. Evans gathered samples of the feed, the water, and the sheep's droppings. He also shared his initial assessment, including observations of the sheep's clinical signs. Dr. Lee then conducted detailed laboratory analysis, identifying the source of the toxin. Her findings enabled Dr. Evans to develop a targeted treatment plan, focusing on counteracting the toxin's effects and supporting the sheep's digestive system. The combined expertise of the two veterinarians was crucial in solving the crisis and saving the flock.

The case of the injured hawk served as another example. The wildlife rehabilitation center received this magnificent bird of prey, injured after hitting a window. Dr. Miller, our avian specialist, immediately assessed the hawk's injuries, noting a fractured wing and internal bleeding. However, the hawk also showed severe dehydration and weakness, along with signs of exhaustion. The challenge here was not just repairing the fractured wing, but ensuring the bird remained stable enough to withstand the surgery without complications. Dr. Again, Chen stepped in, using her expertise to maintain the hawk's vital functions and stabilize it for surgery before the wing repair. Her collaboration with Dr. Miller was key to the hawk's successful recovery. This collaborative approach extended even further, involving the zoology department. They provided dietary advice to help the hawk gain strength during the crucial recovery phase after the surgery was complete.

These were merely snapshots of a larger tapestry of collaboration. Each day brought new challenges and opportunities to collaborate, learn from colleagues, and work together toward a common goal: providing the best possible care for our patients. The success of our practice wasn't simply a matter of individual talent, but a testament to the power of teamwork, the spirit of collaboration, and the strength of a supportive work environment. We are a culture of shared responsibility, mutual respect among colleagues, and a unified pursuit of excellence. The shared purpose and collective ethos of our work make it both challenging and profoundly meaningful. The practice goes beyond the basic provision of care for animals; it delves into a complex web of ethical considerations and responsibilities. This, I understood, was the profession's core: not just work, but a collective pursuit of a common purpose.

Chapter 45: End of Day Reflection: Reviewing the Day's Events

I finally tucked the day's last patient, a fluffy Persian kitten with a persistent cough, into its carrier, heading home with a prescription for antibiotics and a hopeful prognosis. As I locked up the clinic, the quiet hum of the city replaced the usual sounds of barking, meowing, and the rhythmic beeping of monitors. The exhaustion settled in, a familiar weight on my shoulders. But tonight, it felt different. It wasn't just physical fatigue; it was the weariness that clung to the soul, the residue of emotional investment in each case.

I leaned against my car, the cool evening air a welcome contrast to the warmth of the clinic. The day had been a marathon, a relentless succession of emergencies, routine check-ups, and the complexities of chronic care. Each case, each animal, and each interaction with its owners had left its mark. There was the anxious chihuahua, trembling in its owner's arms; the older golden retriever struggling with arthritis; the young cat recovering from surgery–a kaleidoscope of lives intersecting, all demanding attention, empathy, and skill.

The weight of responsibility pressed down; the thought that each life entrusted to my care depended on my knowledge, my judgment, my skill. This wasn't just about technical expertise; it was about being a lifeline, a source of hope and healing in times of distress. The pressure was immense, a constant reminder of life's fragility and the profound responsibility I carried.

But beneath the exhaustion, an unfamiliar emotion stirred — a deep sense of fulfillment. I'd witnessed the resilience of animals, their capacity to heal, and their unwavering trust. I witnessed owners' joyful faces as their pets recovered and the relief in their eyes as anxieties eased. Their witness, healer, and a part of their journey, I was there. This, I realized, was the heart of my profession: the intimate connection forged with animals and their people, the privilege of witnessing both suffering and healing, the satisfaction of making a tangible difference in their lives.

My drive home was usually a blur of exhaustion, a silent retreat into the quiet solitude of my thoughts. Tonight, however, I consciously slowed down. Instead of racing through the familiar streets, I allowed myself to savor the city lights and the cool night breeze against my skin. This deliberate slowing down was a technique I actively incorporated into my life–a way to transition from the intensity of my workday to the peacefulness of my downtime.

At home, instead of immediately collapsing onto the sofa, I attempted to decompress. I began with a warm bath, allowing the soothing warmth to melt away the day's tensions. The scent of lavender essential oil — a deliberate choice for its calming properties — filled the bathroom, a small ritual designed to ease my mind. As the water enveloped me, I allowed my thoughts to wander, not focusing on any problem, but letting my mind be. I found that sometimes the best way to deal with mental exhaustion is to give the mind a break.

After the bath, I prepared a light dinner, something nourishing and comforting. It's a simple act, but it's a vital part of my self-care routine. I don't use unhealthy food as a stress reliever; instead; I focus on preparing a healthy meal and enjoying each bite mindfully. It's a moment of self-nurturing, a quiet act of self-respect.

Later, I took a walk in the nearby park, letting the night air cleanse my lungs and the rhythmic movement soothe my muscles. The darkness, surprisingly, felt peaceful. It offered a contrast to the brightly lit world of the clinic, a reminder of the simplicity and tranquility that existed beyond the scope of my professional life. The quiet sounds of the night–the rustling of leaves, the chirping of crickets–were a balm to my frayed nerves.

This walk was integral to my process of unwinding. It's not merely physical exercise; it's meditation, a chance to reconnect with nature and to detach from the emotional intensity of the workday. During these walks, I often reflect on the day's events, but not analytically or critically. Instead, I allow myself to feel whatever emotions arise–gratitude, exhaustion, frustration, satisfaction. I don't judge these emotions; I acknowledge them, letting them flow through me without resistance.

Returning home, I sat down with a cup of herbal tea, opting for chamomile because of its calming properties. This quiet moment, before bed, was my designated time for reflection. I consciously reviewed the day's events, not to dwell on mistakes or to analyze my performance, but to appreciate the lessons learned and to celebrate the successes. This reflective practice wasn't about self-criticism; it was about acknowledging my efforts, recognizing my achievements, and accepting my limitations.

I started a journal, documenting not only significant cases but also the small, often-overlooked moments. I wrote about the grateful tears in the eyes of an older woman as her dog, after a long illness, took its first tentative steps. Playfully swatting at my hand, a young kitten's mischievous glint in its eye caught my attention. I even wrote about the frustration of dealing with a problematic client, acknowledging the emotions and experiences without judgment.

The journal's purpose was not to be a chronological record; journaling became a crucial element in my self-care strategy, a way to externalize emotions, process experiences, and understand my reactions to them. This reflection, this conscious unpacking of the day's feelings and experiences, was integral to my well-being.

There were inevitable days when the weight of the job felt overwhelming. In those days, I relied on my support system: my friends, family, and colleagues. Talking to them and sharing my anxieties and frustrations was crucial to maintaining my emotional balance. It wasn't about seeking solutions; it was about sharing the burden, acknowledging that I wasn't alone on this journey.

I also incorporated mindfulness practices into my routine, using guided meditation apps to enhance my ability to manage stress. These practices weren't about escaping reality; they were about developing a stronger awareness of my thoughts and emotions, about learning to observe them without judgment, and about fostering a sense of inner calm amid the chaos. Mindfulness allowed me to navigate moments of overwhelming stress or anxiety better, promoting resilience and improving my capacity to deal with stressful situations in the work environment. It transformed the way I approached stressors, helping me to engage with them with a greater sense of calm and detachment.

As I drifted off to sleep, I carried with me not the weight of the day's challenges, but the quiet satisfaction of a job well done, the enduring sense of purpose, and the deep understanding that my work, despite its challenges, was not only meaningful but deeply rewarding. The emotional toll was real, undeniably so. Still, the power of self-reflection, coupled with healthy coping mechanisms, enabled me to navigate the complexities of this emotionally demanding profession, discover inner strength, and fully appreciate the deeply fulfilling aspects of my chosen career. The end-of-day routine wasn't just about unwinding; it was about consciously preparing for the next day, for the next challenge, for the next life entrusted to my care. It was a commitment to holistic well-being, recognizing that to heal others, I must first heal myself.

Chapter 46: The Emotional Toll of Dealing with Loss and Grief

The weight of a tiny, lifeless body in my hands felt heavier than any adult animal I'd ever lost. It was a kitten, barely a few weeks old, brought in by a distraught young girl. Pneumonia, swift and merciless, had stolen its life. The girl's sobs echoed the emptiness in my chest, a familiar ache that settled deep in my bones. This wasn't the first time I'd encountered the stark reality of death in my profession, but it never got easier. Each loss, regardless of the species or age, chipped away at a piece of me, reminding me of the finite nature of life and the fragility of the creatures I had dedicated my life to protecting.

The girl's parents, their own eyes glistening with unshed tears, clutched her close. Their grief mirrored my own; the shared sorrow formed an invisible bond, a silent understanding that transcended words. My carefully crafted words of condolence felt inadequate, hollow against the magnitude of their loss. The professional detachment I often relied on crumbled in the face of such raw, visceral pain. I felt the familiar sting of tears in my own eyes, a testament to the emotional toll this profession had exacted.

Later, alone in my office, the weight of that insignificant life settled heavily on my shoulders. I looked at the empty examination table, still faintly smelling of antiseptic and kitten fur, and the quiet hum of the clinic felt deafening. The girl's cries echoed in the silence, a haunting soundtrack to the grief permeating the air. This was the unspoken truth of veterinary medicine: the constant confrontation with death, the inevitable losses, the profound emotional investment in each animal's life, and the heartache that accompanied their passing.

It wasn't just the animals I mourned. The grief of their owners, their deep connection to their beloved pets, amplified the sorrow tenfold. I had witnessed the unwavering loyalty of a golden retriever to its senior owner, the playful exuberance of a border collie that brought endless joy to a young family, and the comforting presence of a sleek black cat that kept a lonely woman company during her long nights. Their human companions' stories intertwined with those of the animals. To lose one was to lose a part of the other.

One particular memory haunted me. A senior gentleman, Mr. Henderson, had brought his aging Labrador, Buster, in for what would be their last visit. Buster had been a constant companion for over fourteen years, a loyal friend through thick and thin. Mr. Henderson's hands trembled as he stroked the dog's matted fur, his voice choked with emotion as he recounted the memories they had shared. The farewell was agonizingly long; however, the quiet dignity in Mr. Henderson's acceptance was profoundly moving. It reminded me of the profound, unconditional love that existed between humans and their animal companions, a love that transcended the boundaries of species.

The emotional impact of these moments was profound and deeply personal. The unspoken words, the silent understanding, and the shared grief—these were not merely professional interactions; they were intensely human experiences. I bore witness to these life-changing moments, and in doing so, I couldn't help but become emotionally invested in each animal and each human's journey.

My coping mechanisms developed. Setting boundaries between my professional and personal life is something I have learned to maintain. Mindful practices like meditation, deep breathing exercises, and yoga helped me regulate my emotional responses. I understood that to be effective in my role; I needed to prioritize my well-being. Self-care wasn't a luxury; it was a necessity. Without it, the emotional drain of the profession would have overwhelmed me.

I also found solace in writing. I started a detailed journal, chronicling not only the medical aspects of each case but also the emotional nuances, as well as the shared moments of joy and sorrow. The act of writing became catharsis, a means to process my experiences and give voice to the emotions that often went unspoken. It was also a testament to the profound connections formed through my work, a way to preserve the stories of the animals and their people.

My support system also played a crucial role in helping me navigate the emotional challenges that I faced. My colleagues, friends, and family became my lifeline, offering a safe space to share my anxieties and frustrations. They provided a crucial counterbalance to the inherent pressures of my work. These were the people who understood the emotional depths of my profession, who empathized with the emotional toll, and who offered invaluable support and understanding. Their unconditional love and support were a constant source of strength, reinforcing my commitment to the profession I loved.

The emotional toll of veterinary medicine is undeniable — a constant tension between compassion and professional detachment, joy and sorrow, hope and despair. There are days when the weight of loss is overwhelming, when the emotional exhaustion threatens to consume me. There are moments when the fragility of life, the inescapable reality of death, brings me to my knees. However, I've learned that these difficult moments are a part of the journey, a testament to the depth of connection I share with animals and their owners. These experiences have not diminished my passion, but refined it, deepened it, and made it infinitely more profound and meaningful.

I've discovered that healing extends beyond the physical realm. The impact extends to our emotional well-being, our spiritual lives, and also the profound mysteries of existence. This process includes coming to terms with one's grief, commemorating the life of the deceased, and discovering comfort in the collective experience of loss and bereavement. This process involves recognizing the boundaries of my own capabilities, acknowledging my inherent vulnerability, and fully embracing the intricate and often emotionally complex nature of the work that I do. While this profession presents considerable challenges, it offers profoundly rewarding emotional benefits that are just as significant and impactful.

I continue to learn and grow, constantly refining my coping strategies, seeking support when needed, and embracing the richness and complexities of this profoundly rewarding career. Inevitable losses are integral to this journey, shaping my perspective, sharpening my empathy, and deepening my commitment to my chosen to heal art. The emotional toll is real, but so too is the profound sense of fulfillment, the privilege of being a part of these precious lives, and the lasting impact of the connections forged through shared experiences of joy and sorrow. The emotional weight is a constant companion, but it is also a measure of the deep connection to the lives I touch. And in that connection, in that shared journey through life and loss, lies the true heart of veterinary medicine.

Chapter 47: Ethical Dilemmas: Making Difficult Decisions

The quiet hum of the refrigerator in the staff room was a stark contrast to the storm brewing inside me. Dr. Ramirez, a seasoned veterinarian with a reputation for unwavering compassion, had just presented me with a case that would test the very foundation of my ethical compass. A beloved golden retriever, twelve years old and suffering from a debilitating form of bone cancer, lay whimpering softly in a nearby examination room. Its owner, an older woman named Mrs. Gable, was inconsolable, clinging to the hope of a miracle cure. But there was no miracle cure.

Dr. Ramirez's proposal was straightforward: aggressive chemotherapy, a long and arduous process with uncertain outcomes, and a hefty price tag that would strain Mrs. Gable's already limited resources. Despite the low probability of success, he insisted and argued that it was our moral obligation to explore and offer every single treatment option to the patient. He stressed the strong, constant bond between Mrs. Gable and her dog, emphasizing the dog's years of loyal companionship. In his view, refusing her this option, even knowing it wouldn't work, was tantamount to denying her the fundamental right to choose her own course of action. The human-animal bond, he said, is a tapestry woven from loyalty, love, and shared moments.

But my instincts screamed a different message. The chemotherapy, while prolonging life, would likely inflict significant pain and suffering on the already weakened animal. It was a cruel gamble, a roll of the dice against insurmountable odds. The financial burden it would place on Mrs. Gable was a separate but equally significant concern. Could she afford the treatment? Would it ultimately bankrupt her? Was it ethically right to make a creature endure such discomfort to extend its life, especially when that life already suffered excruciating pain?

Ethical considerations formed a tangled Gordian knot, with each thread leading to a different conclusion. Was it my role to always strive for aggressive treatment, regardless of the cost or outcome? Or was it my responsibility to consider the quality of life, the potential for prolonged suffering, and the financial implications of the treatment? The Hippocratic Oath, which I had sworn to uphold, seemed to offer no clear guidance.

I recalled a previous case, a young cat with a severe heart condition. The owners, desperate to save their pet, had pursued every available treatment option, accumulating a substantial debt. Medications masked the cat's pain during its last months, but did not address the underlying condition, causing discomfort. In the end, I wondered if we had truly been kind, if we had made the right decision in pushing forward with aggressive treatments, despite the overwhelming odds and the financial burden on the owners.

The literature on veterinary ethics was abundant, but the application was infinitely complex. Each case was unique, a mosaic of individual circumstances, emotional attachments, and conflicting priorities. The human-animal bond, while incredibly powerful, complicated the matter further. It added layers of emotion and personal investment that were absent in purely human medicine.

With Mrs. Gable and her golden retriever, the situation was further compounded by Mrs. Gable's advanced age and limited financial resources. To recommend aggressive treatment, while potentially fulfilling her wish, would undoubtedly diminish her quality of life and place an undue financial burden on her. The prospect of further impoverishing an older woman, already grieving the impending loss of her beloved pet, felt deeply unjust. To me, that felt more unethical than recommending palliative care.

The conversation with Mrs. Gable was tough. I explained the prognosis, the limitations of the treatment, and the potential side effects associated with it. I also laid out the financial implications, a delicate conversation that required tact and empathy. She cried, of course, but there was a strength in her tears, a sense of understanding. She focused on making her dog's remaining time as comfortable as possible, emphasizing her gratitude for his life and the companionship he had provided.

We opted for palliative care, focusing on pain management and maximizing the dog's comfort. It was a heartbreaking decision, but it felt like the most ethical course of action. It acknowledged the gravity of the situation without subjecting the dog to further unnecessary suffering and financial burden on its owner. The experience reinforced the crucial roles of communication, empathy, and ethical considerations in veterinary medicine.

These were not just medical decisions; they were moral choices, burdened with ethical weight and profound emotional consequences. The gray areas were many, and the boundaries blurred. Yet, within these ambiguities, the compass of compassion must always remain a guiding principle, directing our actions towards the alleviation of suffering and the maximization of quality of life for our patients and their owners. The case also highlighted the importance of understanding the patient's context, including the human-animal bond and the financial realities of the situation.

Navigating these ethical dilemmas became a constant learning process, one that required ongoing reflection, open communication, and a willingness to make hard decisions in the face of uncertainty. The weight of responsibility was heavy, the consequences profound, and the need for ethical clarity absolute. But within the challenges lay an opportunity for growth, a chance to refine my understanding of veterinary medicine, and to deepen my commitment to upholding the highest ethical standards in the practice of healing. The memories, both joyful and heartbreaking, shaped my perspective, refining my understanding of the human-animal bond and reinforcing the importance of providing holistic care.

Later that evening, I sat at my desk, the case still swirling in my mind. I picked up my journal, my faithful companion in processing the emotional complexities of my profession. As I wrote, I realized I was facing ethical dilemmas that weren't isolated incidents, but an intrinsic part of veterinary work. They were inevitable challenges that tested our skills, our empathy, and our moral compass. They forced us to grapple with the inherent limitations of medicine, the inevitability of death, and the complex nature of life itself. These experiences have made me a more thoughtful, compassionate, and competent veterinarian. I learned that true healing transcends the physical, encompassing the emotional and spiritual well-being of both the animal and its human companion. This, I realized, was the valid reward of this profession, the very essence of the vocation I had chosen. It was a journey of continuous learning, growth, and unwavering commitment to the animals in my care, as well as to the people who loved them.

Chapter 48: The Joys of Healing Witnessing Animal Recovery

The sun streamed through the large windows of the clinic, illuminating dust motes dancing in the air. It was a stark contrast to the sterile, clinical atmosphere of the examination rooms, a welcome shift in mood that reflected the overall feeling of the day. Recovering patients' joyful barks and their owners' relieved sighs, not the sterile scent of antiseptic and low whimpers of pain, filled the air today.

It had been an enriching morning. First, there was Pepper, a spry Jack Russell terrier, who, just a week ago, had limped into the clinic with her leg twisted at an unnatural angle. The X-rays revealed a clean break, requiring surgery. Now, she bounced around the waiting room; her tail a blur, her energy a testament to the successful repair. Her owner, a young girl named Lily, hugged Pepper tightly, her face radiant with relief. Watching them, I felt a surge of satisfaction, a deep contentment that transcended the professional accomplishment. It was the purest form of joy, witnessing the restoration of health, the rekindling of a bond, the simple act of healing.

Later, there was Mr. Fitzwilliam's Persian cat, a majestic creature named Midnight, who had arrived with a severe respiratory infection. His labored breathing and listless demeanor had filled me with concern. We'd spent days administering fluids, antibiotics, and supportive care, monitoring his progress closely. Today, however, Midnight purred contentedly on Mr. Fitzwilliam's lap, his breathing now regular and strong. Mr. Fitzwilliam's usually stoic face softened into a gentle smile, his eyes welling up as he stroked Midnight's soft fur. His words, hushed with emotion, were a simple "Thank you," but the heartfelt gratitude conveyed volumes more.

This quiet exchange underscored the profound impact that we as veterinarians have on our clients' lives. It wasn't just about mending broken bones or treating illnesses; it was about mending hearts and restoring hope.

These small victories were the heart of veterinary medicine. They justified the grueling years of study, long hours, emotional strain, sleepless nights, exam anxieties, and the constant pressure to make the right decisions. Over time, they fueled my passion and confirmed my career choice.

The afternoon brought a different joy—the quiet satisfaction that comes with a successful preventive care visit. Mrs. Henderson, an older woman who cherished her aging golden retriever, Buddy, had brought him in for his annual check-up. Buddy, though showing his age with a few gray hairs and a slightly stiffer gait, was otherwise in excellent health. His heart was strong, his teeth were sound, and his spirits were high. Mrs. Henderson's relief was palpable as she heard the report. The simple act of confirming his well-being provided her with an immense sense of peace of mind, a testament to the importance of preventative care in maintaining the quality of life for our beloved animal companions.

Throughout the years, I've witnessed countless such moments—the triumphant return of a once critically ill animal to playful exuberance, the relieved tears of an owner embracing their restored companion, and the quiet confidence instilled by preventive measures. Each case, each recovery, each satisfied owner served as a powerful reminder of why I chose this path, a confirmation of the helpful nature of this unique profession.

It wasn't always easy, of course. The emotional weight of dealing with sickness, injury, and ultimately death was a constant companion. There were days when exhaustion threatened to overwhelm me, when the weight of responsibility pressed down heavily. There were heart-wrenching cases where, despite our best efforts, a patient's condition could not be reversed, and the grief of the owners was palpable. The ethical dilemmas often presented agonizing choices, requiring hard conversations and decisions that tested not just my professional skills but also my empathy and moral compass.

However, even in the face of such challenges, the joys of healing invariably outweighed the burdens. The joy was not merely in the technical skill, the successful surgery, or the effective treatment; it was in the profound connection with the animals and their owners. It was in the tangible evidence of our positive impact on their lives.

I remember one particularly challenging case: someone brought in a young kitten named Whiskers with severe injuries after a car hit him. Bruises and broken bones marred his tiny body. His chances of survival seemed slim; the prognosis uncertain. The kitten's owner, a young girl named Chloe, was inconsolable, her grief echoing the gravity of the situation. The entire team rallied around Whiskers. Days and nights blurred into a cycle of intensive care, meticulous wound care, countless anxious moments, and small, hard-won victories. Slowly, slowly, he improved. He ate again, responded to gentle touch, and eventually, played. It was a painstakingly slow recovery, but every small step forward was a victory. The day Chloe came to collect whiskers, healthy and playful, was one I will never forget. Tears streamed down her face, tears not of sadness but of immense relief and boundless joy. Chloe's profound gratitude touched me deeply, reinforcing the power of the human-animal bond and the profound impact of our work.

This profound bond is at the very core of veterinary practice. It is a relationship of trust, unconditional love, and unwavering loyalty. The owners entrust us with their most beloved companions, relying on our expertise and compassion to ease suffering and restore health. It is a privilege to be a part of this bond, to witness the profound connection, the intense love, and the unwavering devotion shared between humans and animals. The responsibility that comes with this privilege is immense, demanding not just technical proficiency but also unwavering compassion, ethical clarity, and deep empathy.

There are countless stories I could share, many instances where the healing process extended beyond the physical realm, encompassing the emotional and spiritual well-being of both animals and their owners. Each story, each case, each encounter, contributes to the tapestry of my veterinary experience. It is a journey filled with challenges and rewards, heartbreak and joy, but ultimately it is a journey of immeasurable fulfillment.

These are not merely cases; these are relationships. We build connections with each animal, often forming strong bonds with their families. When an animal recovers, it's not just a clinical success; it's a celebration of a strengthened bond between human and animal. The joy we experience is not simply professional; it's deeply personal. It's a testament to the healing power of compassion and the restorative strength of the human-animal bond. This reciprocal empathy, this sharing of joy and sorrow, this mutual reliance—this is the valid reward of veterinary medicine. Veterinarians work in a profession where the line between work and passion blurs, meeting daily challenges with resilience, and experiencing profoundly personal and deeply satisfying victories. This ongoing cycle of challenges, growth, and the sheer joy of witnessing recovery is the very essence of what makes my career so enriching, so rewarding, so fulfilling. It's a vocation that constantly challenges, inspires, and reminds me of the incredible power of healing, both physical and emotional.

Chapter 49: The Importance of Continuing Education: Staying Current

The satisfying clink of the latch on the clinic door signaled the end of another busy day. The lingering scent of antiseptic and the faint echo of contented purrs faded as I locked up, the quiet solitude a welcome change from the constant activity. But even in the evening's stillness, my mind buzzed with the day's events, the intricate details of each case replaying like a film reel in my head. The successful surgery on old Bess, the heartwarming reunion of a rescued puppy with its overjoyed owner, the difficult conversation with a client facing the heartbreaking reality of their pet's terminal illness—each experience, both triumphant and challenging, underscored a fundamental truth: veterinary medicine is a field that demands constant learning, a relentless pursuit of knowledge that never truly ends.

This realization dawned on me early in my career, during those intense years of residency. The sheer volume of information—encompassing the ever-developing understanding of disease processes, advancements in surgical techniques, and the proliferation of new pharmaceuticals and diagnostic tools—was initially overwhelming. It felt like trying to drink from a firehose — a constant struggle to keep pace with the rapidly advancing field. Yet, this very challenge became a source of motivation, a driving force propelling me forward on a journey of continuous learning.

My first brush with the urgency of continued education came during a perplexing case involving a young golden retriever suffering from a mysterious neurological disorder. The initial diagnostics were inconclusive, and despite my best efforts, the dog's condition deteriorated. Feeling the weight of my inadequacy, I reached out to a renowned neurologist at a larger veterinary teaching hospital. His advice, from the newest research and methods, was very helpful. While we ultimately couldn't save the dog, the experience left an indelible mark. It underscored the limitations of my knowledge and highlighted the critical need to stay up-to-date with the latest advancements in the field. The emotional toll of the experience was immense, but it spurred me to make continued professional development a cornerstone of my practice.

From then on, attending conferences, workshops, and seminars became a regular part of my professional life. These events went beyond earning continuing education credits; they offered opportunities to network, exchange ideas, learn from experienced professionals, and stay updated on veterinary advancements. The dynamic discussions and shared experiences proved as valuable as the formal lectures.

The sheer breadth of veterinary medicine also prevents a broad and up-to-date understanding. What I learned in veterinary school formed a solid foundation, of course, but it was merely a starting point. Advancements in diagnostic imaging, particularly MRI and CT scanning, have revolutionized our ability to diagnose and treat a wide range of conditions accurately. Understanding these technologies requires extensive training and continuous learning. Similarly, advancements in surgical techniques, such as minimally invasive surgeries, require rigorous specialized training to master the new skills. The rapidly developing field of veterinary pharmacology requires continuous learning to stay up-to-date on new drugs, their applications, potential side effects, and interactions with other medications.

Online resources, such as professional journals, veterinary-specific websites, and online courses, further expanded my learning opportunities. I learned about the latest research, clinical trials, and best practices using online resources, with no need to go to conferences or workshops. The convenience of having information at my fingertips provided immense flexibility in managing continuing education alongside the demands of a busy practice.

The commitment to lifelong learning extended beyond the formal learning environments. I actively sought mentorship from more experienced veterinarians, observing their techniques, learning from their decisions, and benefiting from their wealth of experience. These relationships proved invaluable, not only for professional development but also for navigating the emotional challenges that inevitably accompany the work. The candid conversations, which involved sharing both successes and failures, created a supportive network that bolstered my professional growth and personal resilience. The exchange of knowledge, insights, and experiences within this network profoundly shaped my approach to veterinary practice.

Ethical considerations within veterinary medicine are constantly developing. Advances in reproductive technologies, the growing awareness of animal welfare, and the ever-changing legal landscape require continuous learning and reflection to ensure ethical and responsible practice. Staying current with the latest guidelines and regulations became an essential component of maintaining a compliant and ethical veterinary practice. The moral dilemmas that arise in this field demand a commitment to continual ethical reflection and learning. It is a practice rooted in empathy, compassion, and a profound understanding of animal welfare, continually shaped and informed by the latest scientific advancements and strengthening societal standards.

Beyond the formal aspects of continuing education, the most valuable lessons often came from personal reflection and self-evaluation. Following each challenging case, I meticulously reviewed my actions, evaluating my approach, analyzing the outcome, and identifying areas for improvement. This introspection, often paired with discussions with mentors or colleagues, proved essential for identifying patterns, correcting mistakes, and refining my skills. This personal reflection — a deeply personal form of continuing education — fostered growth in unexpected ways. It not only honed my technical abilities but also refined my decision-making processes, improving my critical thinking and enhancing my overall professional competence.

The pursuit of continued education is not merely a professional obligation; it's a personal commitment. It's a continuous journey of self-improvement and a testament to the dedication required to excel in veterinary medicine. Challenges marked this journey, but a sense of purpose, passion, and the profound joy of making a difference in the lives of animals and their owners also fueled it.

Continuously pursuing personal and professional development brings immeasurable benefits, fostering competence, confidence, and fulfillment in both my work and personal life. Skilled and compassionate veterinarians stand out because of their unwavering dedication to advancements in this dynamic field. Veterinary professionals must continually develop to meet the ethical challenges, profound healing responsibilities, and developing field. My daily confidence and commitment to excellent patient care come from constantly pursuing knowledge and skills. The joy of making a difference, of alleviating suffering and restoring health, is the ultimate reward, a testament to the power of lifelong learning in veterinary medicine. It's a commitment that fuels my passion, enhances my skills, and strengthens my unwavering dedication to the well-being of animals.

Chapter 50: The Future of Veterinary Medicine: Emerging Trends and Technologies

The quiet hum of the refrigerator in my sparsely furnished office was the only sound competing with the gentle rhythm of my thoughts. The day's challenges, the triumphs, the heartbreaking goodbyes–they all swirled in my mind, a testament to the multifaceted nature of veterinary medicine. But the constant, underlying current was the exhilarating realization that I was working within a field poised for a technological revolution. The future of veterinary medicine was no longer a hazy projection; it was materializing before my eyes, transforming the very fabric of how we diagnose, treat, and understand animal health.

One of the most significant shifts is the increasing integration of artificial intelligence (AI) into various aspects of veterinary practice. Initially, I was hesitant. The thought of AI replacing the human element, the nuanced observation, the intuitive understanding of an animal's subtle cues–it felt almost sacrilegious. However, as I witnessed AI's capabilities firsthand, my apprehension gradually gave way to cautious optimism. AI-powered diagnostic tools, for instance, can analyze medical images with an accuracy exceeding human capabilities many times over. They can detect subtle anomalies in X-rays, ultrasounds, and CT scans, often spotting indicators of disease that might escape the human eye, leading to earlier and more effective interventions.

I recall a case involving a Persian cat with persistent respiratory issues. Traditional diagnostic methods had yielded inconclusive results, leaving us frustrated and the cat suffering. Then, we employed an AI-powered image analysis program on its chest X-rays. The AI detected a minute lung lesion human radiologists had missed entirely. This seemingly insignificant detail, imperceptible to the human eye, led to a targeted diagnosis and treatment plan, ultimately saving the cat's life. This instance highlighted the transformative potential of AI in elevating diagnostic accuracy and efficiency. It allowed us to move beyond the limitations imposed by human fallibility and offered a level of precision previously unattainable.

AI's application extends far beyond diagnostics. AI-driven predictive models that emerge can analyze patient data — breed, age, medical history, lifestyle—to predict the likelihood of future health problems. This lets us take action early, such as changing their lifestyle or giving medicine to prevent illness. This can improve their lives and prevent serious diseases. Because we can predict health problems, we can shift from treating illnesses to preventing them, improving overall animal care.

Using AI also streamlines administrative tasks, freeing up valuable time for direct patient care. AI-powered scheduling systems optimize appointments, reduce wait times, and minimize administrative burdens. This improved efficiency translates to more time spent with patients, fostering stronger relationships among doctors, patients, and owners, and enabling more thorough consultations and examinations. The efficiency gains also enable more manageable workloads, improving the work-life balance of veterinarians, a crucial factor in mitigating burnout.

Beyond AI, the field is experiencing a surge in advancements in telemedicine. Remote consultations enabled by videoconferencing and high-resolution imaging technologies are becoming increasingly commonplace. Telemedicine is beneficial for clients in rural areas with limited access to veterinary care, as well as for managing chronic conditions that require regular monitoring. It extends the reach of veterinary services to underserved communities, bridging geographical barriers and ensuring that all animals have access to necessary care, regardless of their location.

The application of 3D printing in veterinary medicine is another exciting development. 3D printing techniques now create custom-designed prosthetics and implants, resulting in a more precise fit and better functionality. This technology offers a revolutionary approach to orthopedic surgery, enabling the creation of highly individualized devices tailored to each animal's specific anatomy. I remember a case involving a dog with a severe leg injury. A custom-designed implant, created using 3D printing, provided a perfect fit, ensuring a successful recovery. This level of precision and customization wasn't possible with traditional methods.

Advancements in genetic testing are also revolutionizing the approach to animal health. Genetic tests can identify predispositions to certain diseases, allowing for early interventions and personalized management plans. This approach to prevention is important, as it lets us take steps to reduce the chance of illness. This individualized approach reduces the potential for disease development and promotes a healthier life for the animal. It represents a significant shift toward a more preventive and personalized model of veterinary care.

However, integrating these advanced technologies isn't without its challenges. The cost of implementing these new technologies can be prohibitive for some veterinary practices, notably smaller clinics. Addressing this economic disparity is crucial to ensure that all veterinarians, regardless of their practice size or location, have access to these vital tools. We must carefully address the ethical considerations surrounding data privacy and the potential for algorithmic bias. Responsible data management and transparency in algorithmic design are crucial for maintaining ethical standards and protecting patient confidentiality.

Despite these challenges, the future of veterinary medicine is undeniably bright. AI, telemedicine, 3D printing, and genetic testing are converging to create a new era of precision, personalization, and accessibility. Although significant challenges exist, the potential benefits are even more substantial: improved diagnostic accuracy, earlier disease detection, more effective treatment options, and enhanced animal welfare. Continuous learning, ethical reflection, and a collaborative spirit are what the path forward entails. Integrating these advanced technologies will not only transform the way we practice veterinary medicine, but they will also elevate the quality of care we provide to our animal patients, significantly improving their lives and enriching the human-animal bond.

As a veterinarian, the prospect of witnessing and taking part in this transformation is deeply inspiring, a testament to the enduring power of innovation in a field dedicated to healing and compassion. The journey of continuous learning, far from the ending, is just the beginning, leading us into an exciting new era of veterinary medicine. An era marked by collaboration, innovation, and an unwavering commitment to the health and well-being of all animals. The future is not just about technological advancements; it's about leveraging those advancements to foster more compassionate, effective, and accessible veterinary care for every animal, everywhere.

Chapter 51: Developing Essential Skills, Academics and Practical Experience

The crisp scent of antiseptic still clung to my scrubs, a familiar aroma that marked the end of another long day at the clinic. Looking back, the path to becoming a veterinarian wasn't simply a linear progression through lectures and exams; it was a vibrant tapestry woven with threads of hard work, unexpected challenges, and invaluable lessons learned both inside and outside the classroom. The foundation of it all, undeniably, was the meticulous development of essential skills—a blend of rigorous academic pursuit and hands-on practical experience.

My journey started long before the hallowed halls of veterinary school. The veterinary science competition team in high school wasn't just about racing exams and winning trophies; it was a crucible that forged my resilience and honed my practical skills. The pressure in those competitions was immense. I was nervous presenting to the judges, but preparing the diagrams and practicing emergency procedures helped me learn to work well under pressure, think quickly, and prioritize tasks in stressful situations. It was a microcosm of the real-world challenges that awaited me in the veterinary field. We learned teamwork, problem-solving, and the crucial importance of precise execution–skills essential for a successful veterinary career. We dissected specimens, practiced suturing techniques on synthetic skin, and gained an understanding of the intricacies of animal anatomy and physiology. These skills weren't just theoretical knowledge; they were muscle memory, ingrained through countless hours of practice and fueled by a passion for the craft.

College further solidified this foundational knowledge. The intense curriculum of my undergraduate program required a different grit—a stamina for late-night study sessions, a willingness to grasp complex concepts, and an unwavering dedication to mastering the intricate world of animal biology, biochemistry, and anatomy. The academic rigor demanded more than just memorization; it was a period of intense intellectual growth, building a solid base of knowledge that would serve as a cornerstone for my future veterinary career.

Beyond the lectures and exams, actual learning began through hands-on experiences. Volunteer work at local animal shelters broadened my horizons. It wasn't just about cleaning cages and feeding animals; it was about understanding animal behavior, recognizing subtle signs of illness or distress, and building a rapport with creatures from all walks of life. The bond I forged with those shelter animals was immeasurable. The experience taught me patience, empathy, and the importance of clear and effective communication, both with the animals and with their human companions. I quickly learned that successful veterinary care requires more than just clinical skills–it demands a deep understanding of animal behavior and the ability to build trust and confidence with both patients and their owners. It is impossible to separate the healing of an animal from the healing of a community, and it is through this experience that I fully understood this concept.

My summer internships at various veterinary clinics provided an even more profound insight into the realities of veterinary practice. The atmosphere was a whirlwind of activity, a relentless cycle of consultations, examinations, treatments, and surgeries. The hands-on experience in these diverse settings, ranging from large animal practices to small animal clinics, provided me with a truly well-rounded perspective on the profession. Each clinic had its unique rhythm, challenges, and rewards. I assisted with everything from routine vaccinations to complex surgical procedures, learning from experienced veterinarians whose expertise and guidance were invaluable. I recall the undercurrent of nervous energy in the operating room during a delicate surgery, the quiet concentration as we worked together as a team, and the shared relief when the procedure was successful. It was a constant learning curve, a relentless push to refine my skills and absorb as much knowledge as possible. These internships taught me to be resourceful and efficient, to adapt to different environments, and always to prioritize the animals' well-being. It introduced me to the critical importance of communication with clients–explaining complex medical issues in clear and concise language, answering their questions patiently, and providing emotional support during stressful times.

The veterinary school experience itself was transformative. It was more than just an accumulation of knowledge; it was a crucible that tested my resilience, refined my skills, and deepened my understanding of the intricate complexities of animal health. The intensity of the curriculum was unmatched, demanding unwavering dedication and a willingness to embrace challenges. Extensive practical training in the college's state-of-the-art facilities complemented the rigorous academic program. We practiced surgery on simulated models, performed necropsies to hone our diagnostic skills, and engaged in countless hours of hands-on training with real-world patients under the watchful eyes of our professors. These experiences built my technical proficiency, instilling in me a deep understanding of the intricacies of animal anatomy and physiology.

Throughout the entire process, collaboration and teamwork were crucial. Working side by side with fellow students, sharing knowledge and supporting one another through the demanding coursework and clinical experiences, fostered a sense of camaraderie and built lifelong professional relationships. This collaborative spirit wasn't just about academic success; it was about learning from each other, sharing insights, and developing a support network that would prove invaluable throughout our careers.

Beyond the academic achievements and practical experiences, moments of profound personal growth punctuated my journey. The emotional challenges of confronting animal suffering, the heartbreaking goodbyes, and the intense emotional bonds I forged with my patients taught me compassion, empathy, and the importance of emotional resilience. Veterinary medicine isn't just a science; it's an art of compassion, kindness, and an unwavering dedication to healing.

The path to becoming a veterinarian demands a commitment that goes beyond intellectual capacity. It requires grit, resilience, and a profound commitment to animal welfare. Veterinary students face long hours, moments of doubt, and the emotional weight of dealing with illness and loss. However, the rewards are equally profound—the satisfaction of alleviating animal suffering, the joy of witnessing a recovery, and the privilege of forming deep bonds with both animals and their human companions. It's a journey of continuous learning, adaptation, and unwavering dedication. Veterinarians build successful and fulfilling careers by developing essential academic and practical skills. It is a journey worth undertaking, a path that is both challenging and rewarding, leading to a future filled with purpose and meaning. As the learning continues, the skills sharpen, and the passion for healing remains the driving force. The journey is ongoing, and the commitment to animal welfare remains steadfast.

Chapter 52: Building a Strong Support Network: Mentors and Peers

The final year of veterinary school felt like a sprint to the finish line, a blur of late nights fueled by lukewarm coffee and the shared anxieties of my classmates. But even amidst the pressure-cooker environment of exams and clinical rotations, I recognized the vital role my support network played in keeping me grounded and motivated. It wasn't just about academic excellence; it was about the unwavering support of those who believed in me, even when I doubted myself.

One such pillar of strength was Dr. Anya Sharma, my mentor throughout my final clinical rotations. Dr. Sharma, a renowned equine veterinarian, wasn't just a teacher; she was a role model. Her calm demeanor was under pressure, but her unwavering dedication to her patients, and her genuine compassion inspired me profoundly. She had a knack for identifying each student's strengths and weaknesses, providing personalized guidance and tailored advice to help them succeed. She saw potential in me I often overlooked, pushing me beyond my perceived limitations. I vividly remember a challenging case involving a young foal with severe colic. Under Dr. Sharma's expert guidance, I learned to navigate the complexities of the situation, making critical decisions under immense time pressure.

Her post-procedure debriefings were invaluable, not just for the technical aspects, but also for the emotional processing involved in dealing with such critical cases. She taught me the importance of self-reflection, acknowledging both successes and failures as integral parts of the learning process. Beyond her technical expertise, Dr. Sharma's mentorship extended to the emotional aspects of the profession, helping me develop coping mechanisms for the inevitable challenges and stresses inherent in veterinary medicine. Her ability to balance professional excellence with personal well-being served as a guiding principle for my career aspirations. My peer group in veterinary school was equally crucial. The intense pressure of the program forged deep bonds among us, creating a robust support system. We shared late-night study sessions, swapping notes, quizzing each other on complex anatomical diagrams, and offering emotional support when exams felt insurmountable. We celebrated each other's successes, commiserated over failures, and provided a much-needed sense of camaraderie during a demanding period. These weren't just classmates; they became my confidantes, sharing both the joys and the heartbreaks of our shared journey.

We knew the value of collaborative learning, recognizing that the strength of our collective knowledge far outweighed individual efforts. Remember those intense anatomy practical's? We would often stay late, quizzing each other, bouncing ideas off one another, and using each other's strengths to overcome our weaknesses. We fostered a genuine sense of teamwork and mutual respect — a foundation that extended far beyond the confines of the classroom. This sense of mutual support and understanding extends to this day as we continue to navigate the challenges of our chosen profession. Regular calls, sharing of experiences, and the ability to lean on one another during challenging moments have proven to be invaluable components of our individual and collective success.

Beyond the formal mentorship of Dr. Sharma and the close-knit bonds within my peer group, I relied heavily on a broader support network extending to my family and close friends. My parents, despite not having backgrounds in veterinary science, offered unwavering emotional support and encouragement throughout my academic pursuits. Their steadfast belief in my abilities fueled my determination, even during moments of self-doubt. They celebrated my minor victories, offered solace during setbacks, and provided a much-needed sense of stability amidst the whirlwind of my studies. They attended my graduation, their pride palpable, a reminder of the unwavering support that underpinned my entire journey. Their encouragement went further than words. Focusing entirely on studies became possible, as their understanding and empathy eased my anxieties. Their love and faith in me served as the bedrock upon which I built my career.

My circle of friends outside the veterinary school provided a crucial counterbalance to the intensity of my studies. Offering respite from academic pressures, they reminded me of the importance of work-life balance. During moments of frustration and celebration, they listened patiently, offering a much-needed distraction and a different perspective. They would remind me to step back from my studies, ensuring I did not neglect the other aspects of my life. These friendships were anchors, preventing me from losing sight of my well-being amid the demanding nature of veterinary school. They offered perspective, reminding me that my life encompassed more than just veterinary medicine. These friendships were, and remain, vital for my mental well-being, preventing burnout and ensuring a holistic approach to life and career.

Networking extended beyond my immediate circle. Attending veterinary conferences and workshops provided opportunities to connect with experienced professionals from various fields within veterinary medicine. These interactions offered insights into diverse career paths and broadened my understanding of the profession beyond the scope of my academic curriculum. These events also fostered valuable professional connections, establishing a broader network of mentors and colleagues that continues to be a source of support and guidance. The opportunity to observe various approaches to veterinary care, engage in discussions about the latest advancements, and share experiences with peers from diverse backgrounds enriched my understanding and expanded my professional horizons.

Building a strong support network requires effort and ongoing nurturing. Genuine relationships based on trust, respect, and shared goals provide emotional resilience, intellectual stimulation, and guidance. The support of my mentors, peers, family, and friends was crucial not only to my academic success but also to my personal growth and resilience. It was the collective strength of this network that carried me through the challenges and celebrated the triumphs of my journey toward becoming a veterinarian. This interconnected web of support remains vital, constantly strengthening as my career progresses and my relationships deepen. It is a support system I strive to maintain, recognizing its significance in maintaining both professional and personal well-being. The journey to becoming a veterinarian isn't solely an individual pursuit; it's a collaborative effort nurtured and sustained by the invaluable contributions of a dedicated support network. The support provided enabled me to overcome obstacles, grow as a professional, and discover the true meaning of success—a success defined not just by academic achievements, but by the strength of the relationships I forged along the way. Mentorship, networking, and fostering strong, supportive relationships are incredibly important; they form the foundation of a fulfilling and successful veterinary career.

Chapter 53: Managing Stress and Burnout: Self-care and Wellbeing

Entering the professional world quickly tempered the exhilaration of finally graduating. What I had previously romanticized about my profession was swiftly and distressingly replaced by the often painfully honest truths inherent in the daily work. I had underestimated the emotional toll this would take; it was far greater than I had expected. I experienced a profound impact on my mental and emotional health because of the sheer volume of cases, the severity of the illnesses, and the heartbreaking losses that were an inevitable part of the experience. The crushing weight of responsibility for countless animals' lives, coupled with an unrelenting cycle of excessively long working hours, left me in a constant state of struggle and pressure. It was during this period that I realized the crucial importance of self-care, a concept that transcended mere pampering and became a fundamental pillar of my professional survival.

The initial shock of this transition was significant. The combination of urgent situations, high-pressure decision-making, and emotionally difficult interactions with grieving pet owners was more than I could handle. I recall one particularly harrowing night when I worked through a series of critical cases, each demanding immediate attention and critical thinking. A dog suffering from severe internal bleeding, a cat with a fractured leg, and a bird with a respiratory infection all required my immediate attention, and the emotional fatigue of dealing with these issues, one after another, was deeply exhausting. The lack of sleep, coupled with the constant weight of responsibility, took its toll. I became increasingly irritable, experienced difficulty concentrating, and felt emotionally drained. I knew this would not work long-term.

My first step involved acknowledging the problem. I had to admit to myself that I was struggling. This wasn't a sign of weakness; it was a necessary acknowledgment of the demanding nature of the profession. Many veterinarians, especially those early in their careers, often feel immense pressure to perform at the highest level, which can lead to a reluctance to acknowledge their limitations or seek help. This reluctance to acknowledge difficulties is counterproductive and can significantly exacerbate the existing stress. Once I admitted to myself that I was struggling, I could seek solutions.

The next crucial step involved incorporating self-care practices into my daily routine. This wasn't about luxurious spa days; it was about building small, sustainable habits that prioritized my well-being. I started with simple changes: prioritizing adequate sleep, even if it meant sacrificing some study time. I began incorporating regular physical exercise into my schedule, even if it was just a short walk during my lunch break. This simple act of moving my body helped to clear my mind and reduce stress. I also attempted to maintain a healthy diet, consuming nutritious meals to fuel both my body and mind. These minor changes, while seemingly insignificant, had a cumulative effect on my overall well-being. The improvement was gradual, but significant.

Beyond the physical aspects of self-care, I focused on nurturing my emotional and mental well-being. I started practicing mindfulness techniques, such as meditation and deep breathing exercises, to center myself and reduce anxiety. These techniques were instrumental during stressful situations, allowing me to remain calm and focused in the face of challenging cases. I also sought emotional support, either through talking to trusted colleagues, friends, or family. Sharing my struggles with others helped ease some of the emotional burden and provided a sense of perspective. It is crucial to have a supportive network of friends and family to fall back on in times of high stress. One of the most helpful things I discovered was joining a support group specifically for veterinarians. Sharing experiences with others who understood the unique pressures of the profession helped to ease feelings of isolation and provided a safe space to express vulnerabilities without judgment. Knowing that I wasn't alone in my struggles was incredibly validating.

Another significant aspect of self-care involves setting healthy boundaries. I learned to say "no" to additional responsibilities when I felt overwhelmed. This wasn't a selfish act, but a necessary one to preserve my well-being. It was a matter of recognizing my limits and prioritizing my own mental and emotional health. It took time to develop this skill. Initially, I felt guilty about saying no, fearing that I might let others down. But this feeling gradually subsided as I recognized the importance of self-preservation. Learning to say no was a crucial aspect of self-care, allowing me to focus on my current tasks and prevent overcommitment.

Incorporating hobbies and activities outside of work became essential. I rekindled my passion for painting, a hobby I had neglected during veterinary school. This creative outlet provided a welcome distraction from the intense demands of my profession, allowing me to express my emotions and unwind after a long day. I also joined a hiking group, enjoying the beauty of nature and the physical benefits of regular exercise. These activities provided a crucial counterbalance to the intense pressures of my work, offering an opportunity for relaxation and rejuvenation. They served as important reminders that my life extended beyond the confines of the clinic.

Continuing education and professional development also played a vital role in my stress management strategy. Staying up-to-date with the latest advancements in veterinary medicine has helped me feel more confident and competent in my ability. Attending conferences and workshops provided opportunities to network with other professionals, share experiences, and learn new techniques. Continuous learning not only enhanced my professional skills but also boosted my self-esteem and reduced feelings of inadequacy. The sense of ongoing growth and development counteracted the potential for stagnation and burnout.

Time-management techniques were paramount to my success in mitigating stress. Implementing a detailed schedule allowed me to prioritize tasks, allocate time for self-care, and prevent overcommitment. Initially, I resisted using a planner as an unnecessary addition to my already hectic schedule, but it eventually proved invaluable. It allowed me to visualize my workload, identify potential bottlenecks, and strategically allocate time for both professional and personal responsibilities. It helped me structure my day, providing a sense of control over my workload and ultimately leading to a significant reduction in stress levels.

Finally, and perhaps most importantly, I recognized the value of seeking professional help when needed. I initially hesitated to reach out to a therapist, fearing judgment or stigmatization. However, I eventually realized that seeking professional guidance wasn't a sign of weakness but a courageous step towards self-improvement. Therapy provided a safe space for me to process my emotions, develop coping mechanisms, and build resilience. The therapist helped me identify patterns of thinking and behavior that were contributing to my stress and provided me with strategies for managing difficult emotions. This was not a quick fix, but a long-term commitment to self-care that was an ongoing part of my life as a veterinarian. The support and guidance I received helped me overcome challenges and build a sustainable career.

The journey towards managing stress and preventing burnout is not a linear one. It requires constant vigilance, self-reflection, and a commitment to prioritizing self-care. It's a dynamic process, requiring adjustments and adaptations as my life grows. But the effort is undeniably worthwhile. By implementing these strategies, I've been able to transform my relationship with my work. I've learned to appreciate the challenges, find fulfillment in helping animals, and recognize the importance of taking care of myself to care for others effectively. The rewards far outweigh the effort, and it is a continuous journey of self-discovery and improvement that enriches both my professional and personal life.

The path to becoming a successful and fulfilled veterinarian involves not only mastering the technical aspects of the profession but also cultivating resilience and developing strategies to manage the inevitable stress inherent in the work. Self-care, therefore, is not a luxury, but a necessity–a fundamental ingredient in achieving both professional success and long-term well-being.

Chapter 54: Choosing a Specialization, Passion and Career Goals

The crisp white coat felt heavier than usual that day. The weight wasn't just the fabric; it was the weight of expectation, the pressure of impending decisions. Graduation had been a whirlwind of celebrations, but the champagne bubbles had long since fizzled. Now, the daunting reality of choosing a veterinary specialization loomed, a choice that would shape the subsequent decades of my life. It wasn't just about picking a niche; it was about aligning my passions with my career aspirations, a task that felt monumental.

My initial inclination, fueled by childhood dreams and countless hours spent with the family pets, was to pursue small animal practice. The image of comforting anxious pet owners, soothing furry patients with gentle hands, and witnessing the joyous reunions after successful treatments had been a constant visual throughout my years of study. This vision, however, felt simplistic and almost naïve in the light of the complexities I'd witnessed during my internships.

My internship rotations offered invaluable real-world experience beyond textbooks. While I appreciated treating a kitten with a respiratory infection, I became fascinated by equine orthopedics. Observing skilled veterinarians diagnose and treat complex lameness inspired deep respect for their expertise and sparked a new area of interest. The precise procedures, the detailed anatomical knowledge, and the deep bond formed between horse and veterinarian captivated me. I had never considered the level of skill and patience required to handle such magnificent animals.

Equally unexpected was my burgeoning interest in wildlife rehabilitation. A placement at a local wildlife sanctuary broadened my understanding of veterinary medicine to encompass a wider array of species, each with its own unique set of challenges. Working with injured birds, mammals, and reptiles presented unique diagnostic and treatment dilemmas. This experience not only revealed my aptitude for working with a diverse array of species, but also instilled a deep sense of responsibility towards animal conservation. The sanctuary's work opened my eyes to the critical role veterinarians play in maintaining biodiversity and protecting vulnerable populations.

These rotations, while challenging, were incredibly rewarding. They weren't merely practical experiences; they were vital exercises in self-discovery. Each specialization offered a glimpse into a different facet of veterinary medicine, highlighting the unique skill sets, the required personality traits, and the overarching rewards of each path. The diversity of the profession became apparent; the possibilities seemed endless. This realization was both exhilarating and overwhelming.

I created mental checklists: small animal practice - high patient interaction, often emotionally demanding; equine practice - specialized knowledge and demanding physical strength; wildlife rehabilitation - irregular hours, demanding fieldwork, and rewarding conservation work. Each option presented its own unique set of advantages and disadvantages, professional and personal considerations intertwining in a complex tapestry.

In this contemplation, the role of personal values became crystal clear. I realized the 'perfect' specialization wasn't solely about technical skills; my passion for wildlife conservation led me to explore careers that allowed me to make meaningful contributions to environmental protection. The ethical responsibility of practicing veterinary medicine with an emphasis on both animal and human well-being was a paramount value that influenced my decision.

Seeking a more complete understanding, I explored diverse resources that surpassed my own limited experiences. Many conferences and workshops enhanced my professional development in veterinary medicine. To gain a comprehensive understanding of veterinary medicine, I contacted veterinarians from diverse specialties, observed their daily work, engaged in extensive questioning regarding their career trajectories, and sought insights into the practical aspects of their professional lives. Through these invaluable interactions, we gained direct insight into the multifaceted nature of each specialization, encompassing the daily routines, the peaks and valleys of the work, integrating professional and personal life, and the overall sense of fulfillment derived from their chosen careers.

Speaking with established professionals provided me with a wealth of different perspectives, helping me to evaluate the pros and cons of each choice. One equine veterinarian, a woman who exuded quiet confidence and competence, detailed her experiences with surgical procedures on racehorses. Her commitment to her craft was inspiring, but her stories about the relentless demands of the job and the long hours spent away from home also gave me realistic insights into the commitment required. Conversely, a small animal practitioner shared the immense satisfaction of reuniting a lost dog with its emotional owner, a feeling she described as uniquely rewarding. These personal accounts were far more illuminating than any job description could ever be.

As my learning progressed, the differences between the various specializations became increasingly apparent and more easily understood. Each area required a unique and significantly different set of technical skills to perform the job successfully. Diagnosing a heart condition in a canine patient requires a completely different approach than the methods used to diagnose a hoof injury in an equine patient or assess a fractured wing in an avian patient, highlighting the vast differences in diagnostic processes across species and anatomical structures. We observed significant differences in approaches to patient care, encompassing handling techniques and medication administration procedures. The comprehension of these differences, however, proved to be a more involved process than mere skill acquisition, demanding a thorough exploration of the subtleties and underlying principles.

Upon reflecting on my skills and preferences, I realized the importance of maintaining a good work-life balance. While deeply passionate about my work, I also recognized the need for time for my personal life, for hobbies, and for relationships. Some specializations seemed to offer better integration of work and personal life than others. This factor became increasingly important as I considered the long-term implications of my career choice.

Ultimately, the decision to choose a specialization became not only a professional one but a deeply personal journey of self-reflection and honest self-assessment. It was about identifying my strengths and weaknesses, my passions and values, and how they all aligned to contribute to a fulfilling career that was both challenging and rewarding. It was a journey of defining what success truly meant to me–was it financial gain, professional recognition, or the deep sense of satisfaction that came with making a tangible difference in the lives of animals and their owners? The answer, I realized, was a nuanced blend of all three.

My decision-making process wasn't a sudden epiphany, but a gradual unfolding. The more data I collected, the more conversations I had, the more self-reflection I engaged in, and the clearer the path became. It was an iterative process, refining my choices as I gained a deeper understanding of myself and the veterinary profession.

In the end, the decision felt less like a conscious choice and more like the unavoidable and natural result of a long journey of experience, ambition, and personal growth — a culmination of years of shaping oneself and one's path. This wasn't a decision reached in solitude, but a collaborative process that included invaluable input from mentors and colleagues, and crucially, a significant amount of personal consideration. Passion fueled the choice, but a realistic assessment of the profession tempered its intensity; at its heart lay a deep understanding of the true meaning of veterinary practice—not simply healing animals, but acting as a compassionate caregiver, dedicated to the welfare of both creatures and their people. I carefully considered this decision, understanding that it would set me on a path toward a career both fulfilling and meaningful. That decision, which finally provided the clarity I desperately needed, acted as the cornerstone for the next exciting chapter in my veterinary career. Even with all the years I have spent honing my craft, I am still on a continuing journey of refinement, exploration, and further specialization. The world of veterinary medicine is so vast, and it is exhilarating to continue growing within it.

Chapter 55: Giving Back to the Profession: Mentorship and Community Involvement

The realization that my journey was far from over quickly tempered the exhilaration of finally choosing my specialization. Challenging cases and long hours, along with the responsibility of contributing to the wider veterinary community, paved the path ahead. It wasn't enough to excel in my chosen field; I felt a deep-seated urge to give back, to mentor those following in my footsteps, and to take part actively in initiatives that promoted animal welfare and advanced the profession.

My first foray into mentorship was almost accidental. A bright, eager veterinary student, Sarah, shadowed me during her clinical rotations. Her enthusiasm was infectious, her questions insightful, and her dedication unwavering. While I initially treated her presence as a simple learning experience for her, I quickly realized the profound impact she had on me. Her fresh perspective challenged my assumptions, her energy renewed my passion, and her unwavering curiosity forced me to re-examine established practices through a newer lens. We formed a genuine connection, one that went beyond the formal mentor-mentee dynamic.

I shared not just technical expertise but also personal experiences, offering guidance not just on equine orthopedics but also on navigating the complexities of professional life and maintaining a healthy work-life balance. Sarah's eagerness to learn wasn't just about mastering surgical techniques; it was about absorbing the essence of veterinary medicine–the compassion, the dedication, and the unwavering commitment to animal well-being.

This experience underscored the reciprocal nature of mentorship. While I aimed to guide and support Sarah, I discovered I was being mentored by her enthusiasm, her thirst for knowledge, and her unwavering optimism. It was a humbling reminder that sharing knowledge often leads to personal growth and a deeper understanding of one's profession. From then on, I sought opportunities to mentor veterinary students and recent graduates, realizing the profound impact even minor acts of guidance could have. I began volunteering at the local veterinary school, where I conducted workshops on advanced equine surgical techniques and offered one-on-one mentoring sessions to students interested in pursuing a specialization in equine care.

Beyond formal mentorship, I found immense fulfillment in participating in community outreach programs. The local animal shelter relied heavily on volunteer veterinarians, and I readily offered my services. Working alongside other professionals in a less structured setting allowed me to hone a range of skills, including not only technical skills but also communication and teamwork. In stark contrast to the specialized setting of an equine practice, the animal shelter was a melting pot of animal species, varying medical needs and conditions, and a wide range of human interactions, resulting in a vastly different and far broader practical experience. The wide range of cases I encountered, which encompassed everything from providing emergency care for a stray cat with a broken leg to performing routine vaccinations on healthy dogs, significantly broadened my understanding of veterinary medicine and underscored the critical importance of a holistic approach to animal care. Through interactions with the dedicated shelter staff and volunteers, I developed a profound and deeper appreciation for the collaborative and essential efforts that are crucial for advancing animal welfare.

My commitment to community outreach extended beyond the local animal shelter. I took part in several equine rescue operations, assisting in the care and rehabilitation of neglected or abused horses. These experiences were both emotionally challenging and deeply rewarding. Witnessing the transformation of these animals, from traumatized creatures to healthy, confident beings, was a powerful testament to the transformative power of veterinary care. Working with equine rescue organizations also connected me with a network of like-minded professionals and volunteers, fostering collaboration and broadening my understanding of the systemic issues that contribute to animal neglect and abuse. It was a humbling reminder of the crucial role veterinarians play not just in treating individual animals but also in addressing broader societal issues that impact animal welfare.

Continuing education became an integral part of my commitment to giving back to the profession. The rapid advancements in veterinary medicine cause continuous learning and adaptation. I actively participated in conferences, workshops, and online courses, keeping up with the latest research and techniques in equine orthopedics.

Attending these events wasn't just about personal professional development; it was also about engaging with colleagues, sharing experiences, and contributing to the broader dissemination of knowledge within the veterinary community. Presentations at these events allowed me to share my experiences and research findings, fostering a spirit of collaborative learning and inspiring others to pursue continuous professional development.

I realized the value of sharing knowledge through publication. I started documenting my clinical experiences and research findings, and submitting articles to veterinary journals. This endeavor allowed me to reach a wider audience, contributing to the existing body of knowledge in equine medicine. It was a tangible way to give back to the profession, making my insights and expertise accessible to other veterinarians worldwide. Each publication was a testament of my commitment to lifelong learning and the sharing of valuable knowledge for the betterment of animal care.

My commitment to giving back extended to promoting ethical standards within the profession. Taking part in professional organizations provided a platform for addressing critical ethical dilemmas, advocating for animal welfare, and upholding the highest standards of veterinary practice. The discussions and debates within these organizations were invaluable, fostering a deeper understanding of the ethical complexities inherent in veterinary medicine. It was a crucial aspect of my professional journey, reinforcing my commitment to ethical conduct and animal welfare.

These were not just abstract concepts; they were principles that guided every decision, every treatment plan, and every interaction with animals and their owners.

The journey of giving back wasn't merely a matter of professional responsibility; it was a profoundly personal one. It was a continuous process of learning, growth, and self-discovery. Mentoring a budding veterinary professional, rehabilitating a neglected animal, and contributing to scientific literature: these experiences enriched my life far beyond the technical aspects of my profession. The journey served as a potent reminder that the veterinary profession is far more than simply a career path, encompassing a multitude of rewarding experiences and challenges. The commitment I made allowed me to find fulfillment in honing my technical skills and making a meaningful impact on the veterinary profession, the animal kingdom, and those within my local community, resulting in a profound sense of purpose. My contributions and commitments reach far beyond the scope of these particular initiatives, encompassing a much broader range of activities and responsibilities. One's path of learning is an unending journey, a continuous pursuit of knowledge, always motivated by the enduring need to contribute and make a positive impact. It's in that reciprocal exchange of knowledge and compassion where the true heart of veterinary medicine lives.

Chapter 56: Final Reflections: The Journey's Impact

L ooking back, the path from that nervous veterinary science competition participant to the established equine veterinarian I am today feels both incredibly long and surprisingly short. The years blurred into a tapestry woven with threads of late-night study sessions, exhilarating successes, heartbreaking setbacks, and the unwavering support of family, friends, and mentors. It was a journey not just of academic achievement, but of profound personal growth and a deepening understanding of my place within the wider world.

One of the most significant lessons learned was the unexpected power of collaboration. My initial focus was primarily on individual achievement–excelling in exams, mastering surgical techniques, and establishing myself as a skilled practitioner. However, the journey taught me the limitations of a purely individualistic approach. The success of any veterinary practice, whether it be a large animal hospital or a small animal clinic, relies on the cohesive efforts of a team. From the dedicated veterinary technicians and nurses to the administrative staff and, most importantly, the clients, every individual plays a crucial role in providing comprehensive and compassionate care. My experiences with the equine rescue operations, for instance, highlighted the vital importance of collaborative efforts in addressing animal welfare challenges. Volunteers and professionals, with their collective expertise, diverse skills, and unwavering commitment, transformed countless animals' lives, proving that teamwork and shared passion can overcome even the most challenging cases.

Another crucial element of my growth was the continuous learning process. Veterinary medicine is a constantly developing field. New diseases emerge, techniques improve, and research continuously unravels the complexities of animal physiology and pathology. My commitment to attending conferences, taking part in workshops, and engaging with the latest scientific literature wasn't simply a matter of professional obligation; it was a source of immense personal fulfillment. The process of absorbing new knowledge, critically evaluating existing practices, and challenging my assumptions was both intellectually stimulating and deeply rewarding. The act of sharing that knowledge, through publications and presentations, further solidified my commitment to continuous improvement, creating a positive feedback loop of learning and growth.

The mentorship I received and provided helped to shape my journey. I was fortunate to have exceptional mentors throughout my education and career — individuals who provided guidance, encouragement, and unwavering support. Their influence extended beyond technical expertise; mentoring Sarah and other students highlighted the reciprocity inherent in this process. In guiding them, I rediscovered my passion, refined my teaching skills, and gained fresh perspectives on established practices. The experience underscored the importance of cultivating a culture of mentorship, ensuring the continuity of knowledge and compassion within the veterinary profession.

However, the journey wasn't without its challenges. There were moments of self-doubt, periods of intense stress, and occasions when the sheer weight of responsibility felt overwhelming. Sometimes the emotional toll of dealing with sick or injured animals proved too much, requiring reflection, resilience, and a renewed commitment to self-care. These difficulties, though often painful, were invaluable learning experiences. They forced me to confront my limitations, develop coping mechanisms, and seek support when needed. They also instilled in me an empathy that enabled me to understand better the struggles of my colleagues, clients, and even the animals under my care. These experiences taught me the importance of balance — balancing the demands of a demanding career and the need for personal well-being, mental health, and genuine human connection.

The commitment to community outreach was more than just a professional obligation; it was a deeply personal endeavor, a means of giving back to the community that had supported me throughout my journey. The local animal shelter wasn't merely a workplace; it was a crucible of compassion where the transformative power of veterinary care was clear every day. Witnessing the resilience of animals in the face of hardship, the dedication of shelter staff, and the kindness of volunteers solidified my belief in the power of collective action in promoting animal welfare. It was an experience that expanded my understanding of the systemic issues affecting animal health and well-being, highlighting the crucial role veterinarians play not only in individual animal care but also in advocating for larger-scale changes.

Beyond the direct care of animals, taking part in educational programs for the public underscored the importance of responsible pet ownership. Educating community members about responsible pet ownership, animal welfare, and the importance of preventive veterinary care was a profoundly rewarding aspect of my journey. Empowering pet owners with knowledge and resources allowed them to provide better care for their companions, minimizing the need for emergency interventions and fostering stronger human-animal bonds. This commitment to community education further broadened my understanding of the multi-faceted aspects of animal welfare and the interconnectedness of human and animal health.

The ethical dimensions of veterinary medicine were a constant focus. A deep-seated commitment to ethical principles guided the decisions I made, from the diagnostic processes to the treatment plans. This commitment extended beyond individual patient care to include my broader professional responsibilities, encompassing the advancement of the veterinary profession through research, publication, and participation in professional organizations. Active involvement in these professional societies provided a platform to address critical ethical dilemmas, advocate for animal welfare, and uphold the highest standards of veterinary practice. The open dialogue and collaborative efforts within these communities provided a perspective that reshaped the way I viewed things, highlighting the complexities of ethical considerations in veterinary medicine and reinforcing the importance of continuing education and self-reflection in ensuring ethical conduct.

My journey was never just a linear path from student to veterinarian; it was a process of continuous learning, professional growth, and deepening purpose. The challenges I faced, the successes I celebrated, and the lessons I learned shaped me into the veterinarian I am today. This is a story not only of personal accomplishment, but of the profound impact a life dedicated to healing can have.

Learning and the commitment to healing continue without end as the journey goes on. More than successful animal treatment, the ultimate reward is the ongoing journey, the constant pursuit of excellence, and the unwavering commitment to animal welfare and professional growth. The future holds many opportunities to continue this journey, including mentoring the next generation of veterinarians, advancing the field through research and innovation, and contributing to a future where animal welfare is a paramount concern. The dedication, the compassion, and the single-minded pursuit of something just out of reach - these are the hallmarks of a life dedicated to healing, and these are the values I will continue to embrace throughout the unfolding chapters of my life.

Chapter 57: Continuing Passion: The Ongoing Pursuit

The crisp fall air carried the scent of woodsmoke and damp earth, a familiar perfume that always seemed to accompany the quiet reflection that settled over me each fall. The end of another year, another harvest of experiences gleaned from the relentless rhythm of veterinary practice. It wasn't the triumphant climax of a perfectly orchestrated symphony, but the comforting cadence of a well-loved melody. The doctor played each note with practiced ease and deep-seated satisfaction. This wasn't a final curtain, but an intermission, a brief pause before the next act began.

The sheer volume of cases–the colic surgeries performed under the cold light of the operating lamp, the delicate stitching of lacerations, the countless consultations and examinations–blurred together in a mosaic of healing. Each animal is a unique puzzle demanding individual attention and individual solutions. There was the old mare, Bess, whose arthritis had robbed her of her springy gait, the young foal battling a stubborn infection, and the majestic stallion suffering from a mysterious lameness. Each presented a unique challenge, a test of my skills, my patience, and my unwavering commitment to their well-being. These were not just animals; they were individuals with distinct personalities, anxieties, and varying levels of resilience. Each interaction was a lesson in patience, empathy, and the deep bond between humans and animals.

Beyond the clinical aspects, it was the human connection that fueled my ongoing passion for the field. The worried eyes of a distraught owner, the relieved sigh of gratitude upon their pet's recovery–these were the intangible rewards that made the long hours, the demanding work, and the emotional toll entirely worthwhile. The gratitude expressed wasn't just for my skills as a veterinarian, but also for the compassionate care, genuine empathy, and unwavering dedication I provided to both animals and their owners. This was a vocation that transcended the technical aspects of medicine; it was a profound relationship built on trust, shared responsibility, and a deep love for animals.

My involvement in the community remained a steadfast priority. The equine rescue operations continued to thrive, a testament to the collective effort of volunteers, professionals, and dedicated supporters. We had expanded our outreach programs to include educational initiatives within local schools, emphasizing responsible animal ownership and the importance of early veterinary care. The impact on the community was profound, a ripple of compassionate care radiating outward. Through events focused on animal welfare, we raised awareness and funds for vital initiatives, transforming not only the lives of individual animals but also fostering a deeper appreciation for the animal kingdom.

The professional growth continued, too. I embraced new technologies, attended advanced training workshops, and remained an active member of professional organizations. The continuous learning process, the constant striving for improvement, and the quest for new knowledge remained vital aspects of my career, and indeed, my life. My mentor's words echoed in my mind: "The learning never ends; the commitment to excellence continues throughout a career dedicated to healing." His wisdom wasn't merely a professional aspiration; it was a deeply personal philosophy that molded my approach to veterinary medicine and life.

This ongoing commitment to professional development extended to mentoring aspiring veterinarians. I found deep satisfaction in guiding younger practitioners, offering guidance and support, and sharing the lessons I had learned through years of experience. It was a reciprocal process, a feedback loop of teaching and learning, a testament to the enduring power of collaboration and shared purpose. I learned from their enthusiasm, their fresh perspectives, and their innate passion for animal welfare. The dynamic exchange not only enriched their professional trajectories but also mine.

However, the journey wasn't always smooth sailing. There were setbacks, heartbreaks, and losses. The unexpected death of a patient, a difficult diagnosis, or a challenging ethical dilemma -these were moments that tested my resilience, challenged my resolve, and demanded introspection. The capacity to cope with loss, to empathize with the suffering of others (both animals and humans), and to persevere despite hardships were invaluable life lessons. Learning to manage stress, to prioritize self-care, and to seek support when necessary became vital tools in maintaining both my professional and personal well-being.

The ever-growing landscape of veterinary medicine demanded adaptation and innovation. Emerging infectious diseases, the challenges of climate change on animal populations, and the increasing complexity of animal health issues required constant adjustments in my approach. This continuous adaptation, along with my willingness to embrace the unknown, became an essential element of my ongoing professional development. It wasn't merely about responding to new challenges, but about actively shaping the future of veterinary care, engaging in impactful research, and disseminating knowledge to both colleagues and the wider community.

But at the core of it all, the unwavering passion remained–a steadfast beacon illuminating the path ahead. It wasn't just about the technical skills, the procedures, or the scientific knowledge; it was about the profound connection with animals, the deep empathy for their suffering, and the unwavering commitment to their well-being. The bond forged between veterinarian and animal—the unspoken understanding and mutual trust—formed the bedrock of my practice.

The years melted into a tapestry woven with the threads of countless successes, occasional failures, and unending learning. It was a journey of both personal and professional growth, a testament to the power of dedication, resilience, and the unwavering pursuit of excellence. Uncertainty and possibilities lie on the undefined path ahead, yet the commitment to a life of healing and compassionate animal care stays constant. The narrative continues, and each new chapter promises new challenges, fresh insights, and renewed purpose. The journey in itself is the ultimate reward. And as the sun sets on another day, the comforting scent of woodsmoke and damp earth fills the air, a quiet promise of the ongoing pursuit.

Chapter 58: Lessons Learned, Wisdom Gained

The quiet hum of the refrigerator, a constant companion in the night's stillness, was a familiar soundtrack to my late-night reflections. The years stretched before me, a vast and intricate tapestry woven with threads of triumph and tribulation, of joy and sorrow, of unwavering dedication and unexpected setbacks. Looking back, the path wasn't a straight line, but a winding road, full of unexpected turns and breathtaking vistas. I didn't learn those lessons only from textbooks or lecture halls;

One of the most profound lessons was the importance of unwavering commitment. The pursuit of a doctorate in veterinary medicine was a marathon, not a sprint. There were countless late nights fueled by caffeine and sheer willpower, moments of self-doubt that threatened to derail my progress, and challenges that tested my resilience to the limits. Yet, the unwavering commitment to my goal, the unwavering belief in my capabilities, and the steadfast support of my family and friends carried me through. I discovered this commitment wasn't merely about professional achievement; it was about learning to embrace the process, to find joy in the journey, regardless of the outcome. It was the quiet strength that whispers, "Keep going," even when exhaustion threatens to overwhelm. Equally vital was the lesson of embracing failure as a stepping stone to success. Not every surgery was a perfect triumph; not every diagnosis was immediately apparent. Sometimes, despite my best efforts, the outcome did not meet my expectations.

These moments compelled me to confront my limitations, examine my approach, and seek improvement, leading to profound learning. These weren't failures in the truest sense; they were growth opportunities, chances to refine my skills, to deepen my understanding, and to emerge stronger and wiser. The ability to learn from mistakes, analyze setbacks without self-recrimination, and use those experiences to inform future decisions was instrumental in my development as a veterinarian. It was about transforming setbacks into springboards, using failures as fuel for future success.

I was fortunate to learn from exceptional mentors who shared their expertise and demonstrated the power of teamwork and collaboration. My professors, colleagues, and fellow students all contributed to my growth, each interaction providing a unique perspective, fresh insight, and a different approach to problem solving. We didn't limit the collaborative spirit to the academic sphere; the lessons learned through collaboration taught me the value of shared purpose, the power of collective action, and the importance of promoting a supportive environment where individuals could flourish.

The importance of continuous learning was another pivotal lesson. New technologies, diseases, and improved techniques mean that veterinary medicine is constantly growing. The commitment to lifelong learning, to staying abreast of the latest developments, became essential to my practice. Attending conferences, taking part in advanced training programs, and engaging in ongoing research became integral parts of my professional development. The commitment wasn't solely about acquiring new skills; it was about cultivating a mindset of curiosity, a desire to expand one's knowledge, and a willingness to embrace new challenges. This continuous learning process wasn't merely a professional necessity; it became a source of intellectual stimulation and personal fulfillment. It was about cultivating the habit of intellectual curiosity and always striving to improve.

Balancing career goals with personal well-being was a crucial lesson learned. Long hours, intense focus, and emotional resilience are often necessary for the demanding work of veterinary medicine. Prioritizing mental and physical health, self-care became critically important. My professional and personal well-being depended on learning to manage stress, set boundaries, and get support. This wasn't an act of selfishness; maintaining equilibrium demanded mindful exertion, self-understanding, and a dedication to self-protection.

Perhaps the most profound lesson was the discovery of the profound connection between humans and animals. It wasn't simply about treating illnesses and injuries; it was about forging bonds, building trust, and sharing in the joys and sorrows of companionship. The gratitude expressed by pet owners, the unwavering loyalty of animals, and the shared moments of healing became a powerful reminder of the profound interspecies bond. It was in these moments that I witnessed the transformative power of compassionate care, the healing touch of empathy, and the strengthening influence of mutual affection. It was about recognizing the intrinsic worth of every life and the profound connections that bind the human and animal worlds together.

The culmination of all these lessons wasn't a destination, but a journey — an ongoing process of learning, growth, and adaptation. It wasn't about achieving a perfect state of expertise, but about continually striving for excellence, about constantly seeking improvement, and about embracing the challenges and opportunities that lie ahead. The path ahead remains uncertain, full of possibilities and unforeseen hurdles. Still, the commitment remains the same–a life dedicated to healing, a life devoted to compassionate care, and a life enriched by the profound connections with animals and the people who love them. The journey itself, with all its twists and turns, its triumphs and setbacks, is the ultimate reward. As I look towards the future, I carry with me not only the wisdom gained through experience but also an unwavering belief in the transformative power of compassion, dedication, and a deep love for the animal kingdom.

Chapter 59: Inspiring Future Generations: Leaving a Legacy

As dawn broke, painting the sky with streaks of pink and orange, the quiet hum of the refrigerator faded into the background. Now, the years stretched before me, a well-loved map, unlike the blur they once were. The journey wasn't just about earning a doctorate; it was about sowing seeds, nurturing aspirations, and leaving a legacy that would resonate through time, inspiring future generations to embrace the wonders and challenges of veterinary medicine.

A profound sense of purpose filled me, thinking of influencing animal healers. It transcended personal gratification and professional success. It was about creating a ripple effect, a chain reaction of compassion, dedication, and expertise. My mentors — those exceptional individuals who had guided and supported me — had unknowingly laid the groundwork for this ambition. Their unwavering belief in my abilities, their patience in answering countless questions, and their willingness to share their knowledge and experience built my aspirations. Now, it was my turn to pay it forward, to become a beacon of hope and inspiration for those who would follow in my footsteps.

One of the most effective ways to leave a lasting impact, I realized, was through mentorship. Sharing my knowledge, both theoretical and practical, with aspiring veterinarians became a central focus of my work. I started by mentoring veterinary students, guiding them through challenging cases, offering advice during moments of doubt, and celebrating their successes. The energy and enthusiasm of these young minds were infectious, constantly reminding me of the passion that had fueled my journey. Their questions, often insightful and unexpected, challenged my assumptions and continually pushed me to refine my understanding. It wasn't just a one-way street; their eagerness to learn enriched my knowledge and broadened my perspective.

My mentorship extended beyond the formal university setting. I volunteered my time at local veterinary clinics, sharing my expertise with younger colleagues and offering hands-on guidance to veterinary technicians. I found immense satisfaction in witnessing their growth, watching them gain confidence, and mastering new skills. These interactions weren't just about transferring knowledge; they were about fostering a community of support, a network of individuals who shared a common passion and a commitment to animal welfare. Each interaction felt like sowing a seed, planting the hope of a more compassionate and knowledgeable veterinary future.

But formal apprenticeships did not encompass all mentorship. I recognized the power of informal mentorship–the subtle influence of daily interactions, the shared experiences, the unspoken encouragement. Simple acts, such as offering a supportive word during a complicated surgery, taking the time to explain a complex procedure, or celebrating the small victories, all contributed to the cultivation of a supportive environment. We nurtured a culture of learning and collaboration, encouraging questions, viewing mistakes as opportunities for growth, and celebrating success collectively.

Beyond mentorship, I found other avenues to inspire future generations. I became actively involved in public outreach programs, sharing my knowledge and passion with the broader community. Presenting at schools and community events allowed me to connect with young people who might not have considered veterinary medicine as a career path. I shared my story–the challenges, the triumphs, and the unwavering commitment that had characterized my journey. These talks were not just about showcasing the technical aspects of veterinary medicine; they were about fostering a sense of wonder, a love for animals, and a deeper understanding of the interconnectedness of life.

I also leveraged the power of writing to reach a wider audience. Sharing my experiences through articles, blog posts, and eventually, this memoir, allowed me to connect with aspiring veterinarians and animal lovers around the globe. These written narratives offered a window into the world of veterinary medicine, dispelling myths, showcasing the realities of the profession, and inspiring others to pursue their dreams. The act of writing itself became mentorship, a way to share knowledge, offer guidance, and ignite a spark of inspiration in the hearts of others.

My commitment to continuous learning and improvement served as another powerful example for aspiring veterinarians. I actively participated in professional development workshops, attended veterinary conferences, and kept up-to-date on the latest research and advancements in the field. This lifelong pursuit of knowledge wasn't just a professional obligation; it conveyed a powerful message for those I mentored: the importance of continuous growth and the necessity to stay current with the ever-developing landscape of veterinary medicine. I hoped this example would inspire them to embrace lifelong learning as a source of personal and professional fulfillment.

The impact on future generations wasn't solely about imparting technical skills and knowledge. It was also about fostering a deeper understanding of the ethical dimensions of veterinary practice. I emphasized the importance of compassionate care, the significance of respecting animal welfare, and the crucial role veterinarians play in safeguarding the health and well-being of animals and the communities they serve. The ethical foundation of veterinary medicine was a cornerstone of both my teaching and personal practice. The hope was to cultivate a generation of veterinarians who not only possessed technical expertise but also maintained a strong ethical compass committed to compassionate care and responsible stewardship.

I also sought to highlight the importance of collaboration and teamwork within the veterinary profession. Veterinary medicine wasn't a solo pursuit; it was a collective endeavor that required effective communication, shared decision-making, and a willingness to learn from others. My own experiences, both the successes and the challenges, served as powerful illustrations of the benefits of teamwork and collaboration. I encouraged my mentees to value diverse perspectives, to seek the wisdom of others, and to nurture the collaborative spirit that is essential for a thriving profession.

I supported the importance of self-care and work-life balance within the often demanding field of veterinary medicine. Sharing my struggles with burnout and emphasizing the necessity of prioritizing mental and physical health were crucial. The message was that seeking support or prioritizing one's well-being wasn't a sign of weakness; it was an essential element of professional longevity and effectiveness. I hoped that by openly discussing these challenges, I could help future generations of veterinarians navigate the complexities of their careers without compromising their well-being.

The legacy I sought to create extended beyond my direct interactions with aspiring veterinarians. I supported organizations dedicated to advancing veterinary medicine and promoting animal welfare, contributing my time and resources to causes that resonated with my values. This included taking part in fundraising initiatives, volunteering at animal shelters, and advocating for policies that protect animal rights and promote responsible animal ownership. By contributing to these broader initiatives, I aimed to create a lasting impact that would extend beyond my interactions and influence a larger community of animal lovers.

Looking back, the journey hasn't been about reaching a particular destination, but about the continuous process of learning, growth, and adaptation. The path ahead remains uncharted, filled with both the exhilarating anticipation of discoveries and the inevitable challenges of the unknown. But the commitment remains—a life dedicated to healing, a life devoted to compassionate care, and a life enriched by the profound connections with animals and the people who love them. The legacy is not a monument built of stone, but a living legacy carried forward by those my dedication and compassion have touched whose lives. It is the echo of my journey, resonating in the hearts and minds of future generations of animal healers committed to a world where every creature receives the compassionate care it deserves. The journey continues; the legacy endures.

Chapter 60: The Enduring Power of Compassion: A Legacy of Healing

The sun dipped below the horizon, casting long shadows across the veterinary clinic as I prepared to leave for the day. The day had been long, filled with the usual mix of triumphs and challenges: a successful surgery on a beloved family pet, the heartbreaking euthanasia of an aging animal companion, the comforting words offered to distraught owners. Each encounter, each interaction, each shared moment of joy or sorrow, underscored the profound and enduring power of compassion in veterinary medicine. It was more than a mere skill, technique, or professional requirement; it was the invisible thread that wove together all aspects of my life as a veterinarian.

I realized that compassion extended beyond technical healing to encompass the full spectrum of human and animal interaction. It meant listening and supporting a grieving client, gently reassuring a frightened animal to build trust, and tailoring each treatment plan to the patient's physical and emotional needs. It also involved collaborating with colleagues, sharing knowledge, and supporting one another, knowing that teamwork often leads to the best outcomes.

The healing process, I learned, rarely existed in isolation. It frequently involved a delicate dance between the physical and the emotional, between the animal and its human companions. The bond between an animal and its owner was often profound and irreplaceable. Witnessing that bond strengthened my commitment to providing not only excellent veterinary care but also compassionate support to the people who entrusted their beloved animals to my care. It was a holistic approach to healing that encompassed the physical, emotional, and spiritual well-being of both the animal and its human companion.

My understanding of compassion deepened over the years through countless encounters with animals and their owners. There was an older golden retriever, diagnosed with a terminal illness, whose owner, a stoic and reserved woman, broke down in tears at the prognosis. At that moment, I didn't just offer medical advice; I offered empathy, acknowledging her grief and validating her emotions. Through gentle words and a compassionate touch, I helped her navigate the emotional complexities of her beloved pet's impending death. I discussed options with her; It was a profound experience that underscored the interconnectedness of animal and human emotions in the healing process.

Or there was the young boy, no older than eight, who brought his injured cat to the clinic. His concern for his feline friend was palpable. I treated him with respect and included him in the examination process, explaining everything in terms he could understand. The act of involving him fostered a sense of control and trust, lessening his anxieties. His cat's recovery became a journey we shared, demonstrating the healing power of participation and shared responsibility.

These encounters served as constant reminders that veterinary medicine wasn't simply about diagnosing and treating illnesses; it was about forging connections, building relationships, and understanding the profound emotional bonds between humans and their animal companions. It was about recognizing the emotional toll that illness, injury, or loss could take on families and offering support and understanding during challenging times. My role extended beyond the technical skills of my profession; it encompassed the role of a compassionate listener, a supportive friend, and a guide during times of uncertainty and grief.

The story of healing extended beyond the clinic walls. Through my involvement in community outreach programs, I witnessed firsthand the transformative power of compassion in the broader community. Taking part in animal welfare initiatives, volunteering at shelters, and working with rescue organizations revealed the significant impact compassion could have on a larger scale, inspiring others to join the cause. It showed me that compassion wasn't merely an individual act but a collective force capable of making a profound difference in the lives of animals and the people who cared for them.

My dedication to compassionate care extended beyond individual animals and their owners, encompassing an unwavering commitment to promoting animal welfare on a broader scale. Advocating for responsible pet ownership, supporting legislation that protects animals from cruelty, and promoting humane treatment were integral to my professional and personal philosophy. I believed that compassion should guide every aspect of our interactions with animals, from our relationships to the broader policies and practices that shape our society. This commitment was not simply a component of my professional life; it permeated my values and guided my daily actions.

The pursuit of continuous learning and development played a crucial role in strengthening my capacity for compassion. With each new skill I gained and each piece of knowledge I gained, my ability to provide better, more informed, and compassionate care expanded. Staying up-to-date with the latest advancements in veterinary medicine, engaging in professional development workshops, and collaborating with colleagues has enriched my practice, enabling me to provide even more effective and empathetic care. This ongoing pursuit of knowledge and skills was not just a professional obligation; it was a necessary step towards cultivating a more compassionate and effective veterinarian.

The importance of teamwork and collaboration within the veterinary profession also played a role in my ability to provide compassionate care. Recognizing that I couldn't possibly possess all the knowledge or expertise needed to treat every case, I embraced a collaborative approach, seeking advice from colleagues, consulting with specialists, and working together with other veterinary professionals to ensure the best possible outcomes for my patients. This collaborative spirit not only enhanced my capacity for compassionate care but also fostered a more supportive and enriching work environment.

Taking time for myself, prioritizing my own mental and physical well-being, and engaging in activities that brought me joy became essential components of my work-life balance. These activities weren't indulgences; they were necessary practices that helped me maintain my energy, focus, and capacity for empathy.

I didn't define my legacy by awards or achievements; instead, I focused on teaching empathy, kindness, and unwavering dedication to animal well-being.

Acknowledgments

To begin with, I am very thankful to all the animal enthusiasts who inspired this narrative. Your constant passion and dedication to these amazing animals never cease to fill me with both wonder and admiration. For the future veterinarians who are dreaming of establishing careers dedicated to working with and helping animals, a talented group of young artists deserves special recognition for their vibrant paintings that have brought Dr. Samantha Green's story to life. Her creative vision uniquely and captivatingly shaped the narrative, thus providing an extra layer of depth to the story.

To my family, whose unwavering support and encouragement made this book possible. Thank you for believing in my dreams and sharing in the journey. Your love and faith have been my guiding light, and I am forever grateful for your presence. You have been my rock, my cheerleader, and my inspiration. This book is as much yours as it is mine.

I'd also like to thank my illustrator, Karen Shayler, and my editor, Roxana Coumans, for their indispensable help and support during the book's publication. This story wouldn't be alive without your keen eye, insightful feedback, and endless patience. This project has benefited from your dedication to excellence and belief in its success.

I am also expressing my gratitude for the invaluable guidance and expertise that my reviewers have provided, particularly including my mother, Billie Aylesworth of Diamond B Cutting Horses, and extending a special thank you to Dr. Jennifer Boucher of Diamond J Veterinary Services. The story has gained authenticity and depth, which directly results from your valuable feedback and insightful contributions. Your love for animals, as well as your commitment to excellence, consistently serves as a source of inspiration for me.

We want to express our appreciation to every reader who has begun this journey with Dr. Samantha Green for their support and enthusiasm. The existence of this book directly results from your passion for stories, besides your faith in the magic that comes from friendship and the act of never giving up. We sincerely hope that this narrative will serve as an inspiration for you to pursue your goals, cultivate your friendships, and treasure the relationships that enhance your existence.

About the Author

Brett Shayler, an author, has been passionate about horses and the natural world his entire life. Raised on a ranch surrounded by diverse animals, his storytelling vividly reflects the profound and enduring connections between people and the animals they cherish and care for throughout their lives. Drawing inspiration from his family's vast acreage and the wild horses that roam freely upon it, Brett lovingly crafts heartwarming tales that not only celebrate the bonds of friendship but also extol the virtues of a deep respect for the natural world and the remarkable connections that exist in the relationships between people and animals. Through his books, he hopes to cultivate curiosity, compassion, and a thirst for knowledge in young readers. Brett often escapes to nature, hiking and observing wildlife, finding inspiration for his writing away from his desk. Brett proudly holds membership in both the prestigious National Cutting Horse Association and the equally esteemed American Quarter Horse Association, demonstrating his deep commitment to these organizations and the equestrian world.

www.ingramcontent.com/pod-product-compliance
Lightning Source LLC
Chambersburg PA
CBHW022007050726
47499CB00003BA/695